Emma Järvinen

Investigations of Angular Light Scattering by Complex Atmospheric Particles

Investigations of Angular Light Scattering by Complex Atmospheric Particles

by
Emma Järvinen

KIT Scientific Publishing

Dissertation, Karlsruher Institut für Technologie (KIT)
Fakultät für Physik

Tag der mündlichen Prüfung: 20. Mai 2016
Referenten: Prof. Dr. Thomas Leisner, Prof. Dr. Manfred Wendisch

Impressum

KIT Scientific Publishing

Karlsruher Institut für Technologie (KIT)
KIT Scientific Publishing
Straße am Forum 2
D-76131 Karlsruhe

KIT Scientific Publishing is a registered trademark of Karlsruhe Institute of Technology. Reprint using the book cover is not allowed.

www.ksp.kit.edu

Print on Demand 2016

ISBN 978-3-7315-0554-9
DOI 10.5445/KSP/1000056601

Investigations of Angular Light Scattering by Complex Atmospheric Particles

Zur Erlangung des akademischen Grades eines
Doktors der Naturwissenschaften
(Dr. rer. nat.)

bei der Fakultät für Physik
des Karlsruher Instituts für Technologie (KIT)

genehmigte
DISSERTATION

von

M.Sc. Emma Järvinen
aus Vantaa

Datum der mündlichen Prüfung: 20. May 2016
Referent: Prof. Dr. Thomas Leisner
Korreferent: Prof. Dr. Manfred Wendisch

Tiivistelmä

Epäsäännöllisiä hiukkasia tavataan kaikkialla ilmakehässä. Niitä voi löytää kaikenkokoisina ja -muotoisina, ja niillä on eri tason monimuotoisuutta. Ilmakehän hiukkasten suuren kirjon takia niiden valonsirontaominaisuuksien ennustaminen on haastavaa, varsinkin kun epäsäännöllisille hiukkasille ei ole olemassa analyyttistä ratkaisua, joka suoraan ratkaisisi elektromagneettisen aaltoyhtälön. Kuitenkin tämän haasteen ratkaiseminen on tärkeää, sillä nämä hiukkaset vaikuttavan maapallon energiataseeseen sirottamalla ja absorboimalla auringonsäteilyä ja osaltaan lisäävät ilmastoennusteiden epävarmuutta. Siksi yhä paljon vaivannäköä käytetään optiikan perustutkimukseen, jossa teoreettisesti, numereerisesti tai kokeellisesti yritetään ymmärtää epäsäännöllisten hiukkasten optisia ominaisuuksia. Tässä tohtoritutkinnossa olen ottanut kokeellisen lähestymistavan ja tutkin ilmakehän hiukkasten kulmariippuvaisia sirontaominaisuuksia alan johtavilla optisilla mittausmenetelmillä hiukkaskammiokokeissa ja lentokonemittauksissa tutkimuslentokoneilla, kuten saksalaisella HALO (High Altitude and LOng Range) -mittauslentokoneella. Hiukkaskammiokokeet osoittivat, että ilmakehän hiukkasten depolarisaatio-ominaisuudet lähellä takaisinsironta kulmaa ovat suuresti riippuvaisia hiukkasten koosta, kun taas pienet muutokset hiukkasten morfologiassa eivät olleet havaittavissa, jos hiukkasten perustava muoto erosi pallosta. Toisaalta nämä mittaukset olivat herkkiä havaitsemaan muutoksia homogeenisesta pallomaisesta muodosta, mikä osoitettiin kokeilla, joissa havaittiin α-pineenihiukkasten lasimainen tai viskoosinen olomuoto depolarisaatiosignaalilla. Sitä vastoin eteenpäin sironneen valon diffraktiokuvion analyysi tarjosi tietoa mikrometriluokan kokoisten hiukkasten mikrorakenteesta. Tätä menetelmää sovellettiin laboratoriokokeissa ja kenttämittauksissa jäähiukkasten ominaisuuksien tutkimiseen. Nämä tulokset antoivat suoraa näyttöä, että jäähiukkasten muodostuminen ja kasvu rosoisiksi tai epäsäännöllisiksi on tavallista ilmakehässä, kuten ovat osoittaneet aiemmat epäsuorat havainnot, kuten satelliittimittaukset. Jääkiteiden epämuodostumat johtivat sironneen valon kulmalliseen jakautumiseen, joka oli muuttumaton mittauspaikasta toiseen, ja vaihefunktioon, jolla oli matala epäsymmetriaparametri, 0.796 ± 0.02. Nämä havainnot tukevat alan viimeisintä kehitystä, joka painottaa optisissa malleissa rosoisuuden tai monimutkaisuuden lisäämistä mallijäähiukkasiin. On myös odotettavissa, että havaitut valon kulmariippuvaiset sirontaominaisuudet voivat vaikuttaa ratkaisevasti ilmastoennusteisiin.

Zusammenfassung

Irreguläre Teilchen sind in der Atmosphäre sehr häufig anzutreffen. Sie sind in allen Größen vorhanden und haben unterschiedliche Formen und Komplexitätsgrade, von irregulären Staubpartikeln bis hin zu größeren, facettierten Eispartikeln. Mit einer derart riesigen Auswahl von Formen, Größen und Mikrostrukturen ist die Vorhersage ihrer Lichtstreueigenschaften eine große Herausforderung. Die Überwindung dieser Herausforderung ist aber essenziell, da diese Partikel den Energiehaushalt der Erde durch Streuung und Absorption beeinflussen und zu Unsicherheiten in Klimaprognosen führen. Die Berechnung der Lichtstreueigenschaften komplexer Partikel ist nicht trivial, da es dafür keine analytische Lösung gibt. Nach wie vor wird viel Aufwand darin investiert, das Problem durch theoretische, numerische oder experimentelle Arbeiten zu lösen. In der vorliegenden Doktorarbeit habe ich einen experimentellen Ansatz vorgenommen, um das Lichtstreuverhalten von komplexen Partikeln mit neuesten optischen Methoden in Wolkenkammerexperimenten und mit Messflugzeugen, wie z.B. dem deutschen HALO (High Altitude and LOng Range) Messflugzeug, zu untersuchen. In den Wolkenkammerexperimenten konnte ich nachweisen, dass die Rückstreu-Depolarisationeigenschaften von atmosphärischen Partikeln im Wesentlichen von der Größe der Partikel abhängig sind. Kleine Änderungen im Komplexitätgrad der Partikel zeigten allerdings keine signifikante Wirkung. Andererseits waren die Depolarisationmessungen sehr empfindlich auf kleine Abweichungen von der sphärischen Partikelform, was den Nachweis eines zähflüssigen Zustands im Falle von sekundär-organischen Aerosolpartikeln ermöglichte. Messungen der räumlichen Verteilung von vorwärts gestreutem Licht enthält hingegen Informationen über die Mikrostruktur von größeren Partikeln. Diese Methode habe ich in Labor- und Feldmessungen angewandt, um den Grad der Eispartikelkomplexität zu bestimmen. In direkten insitu Messungen wurde nachgewiesen, dass Komplexität eine weit verbreitete Eigenschaft von atmosphärischen Eispartikeln ist, was schon indirekt, durch Satellitenbeobachtungen indiziert war. Die Eispartikelkomplexität resultiert in einer Winkelstreufunktion mit einem niedrigen Asymmetrieparameter von 0.796 ± 0.02 und zeigt wenig Variation zwischen verschiedenen Messorten in der Atmosphäre. Diese Resultate unterstützen die momentane Entwicklung in der Eispartikelmodellierung, in der Kompletxität als ein fester Bestandteil angenommen wird. Außerdem könnte der niedrige und stabile Asymmetrieparameter deutliche Konsequenzen für die Klimaprognosen haben.

Abstract

Irregular particles are ubiquitous in the Earth's atmosphere. They are found in all size ranges, consist of a variety of shapes and have different degrees of complexity; from sub-micron irregular mineral dust particles to several micrometer faceted ice crystals. With such a variety of shapes, sizes and microstructures predicting their light scattering properties is a challenge. Overcoming this challenge is an important task since these particles influence Earth's energy budget by scattering and absorbing sun light and, thus, add to the uncertainties in climate predictions. Yet, solving the light scattering problematic of complex bodies is not straightforward, as no analytical solution for non-spherical or non-isotropic particles is available. Still intense research focus is invested in understanding the light scattering by complex particles through theoretical, numerical and experimental work. In this thesis I have taken an experimental approach to study the angular light scattering properties of complex atmospheric particles by deploying state-of-the-art optical methods in cloud chamber studies and onboard research aircrafts, like the German High Altitude and LOng Range Research Aircraft (HALO). In the cloud chamber experiments it was found that the near-backscattering depolarisation properties of atmospheric particles were largely dependent on the particle size, whereas small changes in the particle morphology could not be detected if the underlying shape was non-spherical. On the other hand, the near-backscattering depolarisation measurements were extremely sensitive to small changes from the spherical shape, which was demonstrated by observing the viscous phase state of secondary organic α-pinene particles. On the contrary, spatially resolved forward scattering measurements revealed information of the particle microstructure in the case of super-micron particles. This method was used to observe and quantify ice particle structural complexity, i.e. deviation from the pristine shape, in laboratory and in field experiments. The obtained results showed direct evidence that ice particle complexity is a common feature in natural ice clouds as indicated previously by indirect methods, such as satellite observations. The high degree of ice particle complexity led to a scattering phase function with a low asymmetry parameter of 0.796 ± 0.02 that was observed to be stable from one measurement location to another. These results support the recent developments in the atmospheric ice particle optical modelling, which have inferred that complexity is needed in particle models to explain the observations. It is also anticipated that the observed stable and relatively low asymmetry parameter can have significant consequences for predicting the cloud radiative effect.

List of Notations

ACRIDICON-CHUVA	Aerosol-cloud-precipitation interactions and their thermodynamic, dynamic, and radiative effects by in-situ and remote sensing measurements over Amazonia
AD	Analog-to-digital
AERONET	The Aerosol Robotic Network
AIDA	Aerosol Interaction and Dynamics in the Atmosphere
ATD	Arizona Test dust
AWI	Alfred Wegener Institute
BE	Beam expander
BMBF	Federal Ministry of Education and Research
CALIOP	Cloud-Aerosol Lidar with Orthogonal Polarisation
CALIPSO	Cloud-Aerosol Lidar and Infrared Pathfinder Satellite Observation
CDP	Cloud Droplet Probe
CERN	The European Organisation for Nuclear Research
CIP	Cloud Imaging Probe
CLOUD	The Cosmics Leaving Outdoor Droplets experiment
CLOUD8	The 8th CERN CLOUD campaign that took place from September to December in 2013
CLOUD9	The 9th CERN CLOUD campaign that took place from September to December in 2014
CoCiP	The contrail cirrus prediction model
CPC	Condensation Particle Counter
CPI	Cloud Particle Imager/Imaging
Δ_L	Degree of linear depolarisation
δ_L	Particle linear depolarisation ratio
δ_C	Particle circular depolarisation ratio
DDA	Discrete Dipole Approximation
DFG	The German Research Foundation
ECMWF	European Centre for Medium-Range Weather Forecasts

FOV	Field of view
g	Asymmetry parameter
GLCM	Grey-level co-occurrence matrix
GMD	Geometric mean diameter
GSD	Geometric standard deviation
GTP	Glen-Taylor prism
HALO	High Altitude and LOng Range Research Aircraft
HGF	Helmholtz Association
HR-ToF AMS	A high-resolution time-of-flight aerosol mass spectrometer
IDL	Interactive Data Language
IWC	Ice water content
LIDAR	Light Detection And Ranging
LPR	Liquid crystal polarisation rotator
LVR	Liquid crystal variable retarder
LOD	Limit of detection for depolarisation ratio
MAPMT	Multi-anode photomultiplier array
MBW	A chilled mirror dew point hygrometer
ML-CIRRUS	The Mid-Latitude CIRRUS experiment
NA	Numerical aperture
O/C	The atomic oxygen to carbon ratio
PARASOL	Polarization & Anisotropy of Reflectances for Atmospheric Sciences coupled with Observations from a Lidar
PHIPS-HALO	The Particle Habit Imaging and Polar Scattering Probe
PIP	Precipitation Imaging Probe
PPD-2K	The Particle Phase Discriminator mark 2, Karlsruhe Edition
PN	Polar Nephelometer from University Blaise Pascal
PMMA	Poly(methyl methacrylate)
PTR-TOF	The Proton Transfer Reaction Time of Flight Mass Spectrometer
PÖLY01	AIDA campaign
RACEPAC	Radiation-Aerosol-Cloud Experiment in the Arctic Circle
RH	Relative humidity
RICE03	Third AIDA campaign in the series Rough ICE
SA	Sulphuric acid
SEM	Scanning Electron Microscope

SID-3	The Small Ice Detector mark 3
SIMONE	The Scattering and Depolarisation Instrument
SMPS	The Scattering Mobility Particle Sizer
SOA	Secondary organic aerosol
TDL	A tuneable diode laser
UHSAS	An Ultra-High Sensitivity Aerosol Spectrometer
UT/LS	Upper troposphere and lower stratosphere
VMD	Volume median diameter
WCB	Warm conveyor belt
WP	Wollaston prism

List of Variables

Variable	Unit	Description
A	m^2	Particle projected area
ar	–	Particle area ratio
α	$^\circ$	Half opening angle of a detector
$\vec{B}$	T	Magnetic field
C, C'	–	Complexity parameter
C_s	m^2	Scattering cross section
$\frac{dC_s}{d\Omega}$	$\frac{m^2}{sr}$	Differential scattering cross section
cf	–	Photon multiplier gain correction factor
$\delta_{p,h}$	–	Particle linear depolarisation ratio measured with parallel incident polarisation
$\delta_{p,v}$	–	Particle linear depolarisation ratio measured with perpendicular incident polarisation
d	m or µm	Particle diameter
$\vec{E}$	$V\,m^{-1}$	Electric field
$\hat{\mathbf{e}}_{\perp,\parallel}$	–	Unit vector
ϵ_0	$F\,m^{-1}$	Permittivity of free space
g	–	Asymmetry parameter
i	–	$\sqrt{-1}$
I	–	Stokes parameter
I_0	W	Laser intensity
$\vec{J}$	$A\,m^{-2}$	Electric current density
$\vec{k}$	–	Constant wave vector
k	–	Wave number
k_e	–	Complexity parameter
Ω	sr	Solid angle
ω	s^{-1}	Angular frequency
μ_0	$H\,m^{-1}$	Permeability of free space
P_L	–	Degree of linear polarisation

P_C	–	Degree of circular polarisation
P	m	Particle perimeter
p	–	Scattering phase function
Q	–	Stokes parameter
R	m	Radius
r	m	Distance from the scattering center to detector
$\vec{r}$	–	Position vector
ρ	$\mathrm{C\,m^{-3}}$	Electric charge density
σ	–	Standard deviation or complexity parameter
S_{mn}	–	Scattering matrix element
U	–	Stokes parameter
V	–	Stokes parameter
x	–	Size parameter
ζ_v^{acw}	ppm	The available condensable water vapour mixing ratio

Contents

1. Introduction

We have all seen a rainbow after a rainy day or a halo around the Sun at the presence of cirrus clouds. We know that the sky is blue and that the sunset shows vigorous colours of red. We have seen images of sand storms or wondered about the yellowish smog during the Beijing Olympic games. These phenomena are all postulations of light interactions with atmospheric particles: rain droplets, ice crystals or large aerosol particles. The theory of light scattering has puzzled the humankind through ages. It was Empedocles, in the fifth century BC, who asked, why do we see objects. He postulated that human eye is capable of emitting a light beam that is then reflected from an object back to the eye. Naturally, the theory failed to explain, why humans cannot see in the dark. Later, Euclid described light as a traveling straight beam. He wrote the opus *Optica* that described the laws of reflection mathematically.

Long has past since the days of Empedocles and Euclid. Today we know that light is not a straight beam but electromagnetic waves traveling in a medium. Yet, in the modern world, understanding the principles of light scattering and applying them to everyday problems has not lost its importance. It is in this era that we are facing global problems that require scientists from several fields to work together to solve them. One of these problems is the ongoing climate change caused by global warming of the atmosphere.

Without our atmosphere, there would hardly be life on this planet. It blocks some of the Sun's hazardous rays from penetrating to Earth, it is a reservoir of the oxygen we breath and it keeps us warm. It can be seen as a thin blanket covering the Earth; it absorbs sunlight, emits thermal radiation back to Earth's surface and prevents thermal radiation from the surface from escaping to the space. Without it, the mean temperature of the Earth would be a mere -18°C; only some degrees higher than the mean temperature on the Moon.

This so-called *greenhouse effect* maintains our planet liveable. However, the drastic increase in the emissions of greenhouse gases due to human activities since the industrial revolution has changed the Earth's *energy budget*. To maintain the energy balance, the Earth's climate has responded by increasing the mean temperature. The magnitude of this change can be assessed by defining the *radiative forcing* that tells us the change in the net irradiance (solar plus long-wave; in $\mathrm{W\,m^{-2}}$) at the tropopause due to an externally imposed perturbation (Ramaswamy et al., 2001). For determining the radiative forcing, it is equally important to understand both the influence of the external perturbation, as well as the pre-industrial or unperturbed energy budget, as the radiative forcing is always the difference of these two.

From all the atmospheric constituents, cloud and aerosol particles create the largest uncertainties to estimates and interpretations of the Earth's changing energy budget (Boucher et al., 2013). Therefore, for accurate predictions of our future climate it is essential to understand the role of the atmospheric particles. These particles can affect Earth's energy budget directly by scattering and absorbing shortwave and long-wave radiation or indirectly through aerosol-cloud interactions. In order to properly evaluate the direct climate effect of atmospheric particles, we need to have a good understanding of their light scattering properties. This is where atmospheric particle optics come into play. However, it is not an easy task to tackle the interactions between light and atmospheric particles. Let's face it: if one excludes cloud droplets and aqueous aerosol particles, all the other atmospheric particles are more or less complex, non-uniform, have roughened surfaces or can undergo phase changes. There is no light scattering theory that can analytically solve the light scattering by these complex particles. And this is just one side of the coin.

"Light scattering by atmospheric particles - a two-way problem"

It was Bohren and Huffman (Bohren and Huffman, 2008) who described the light scattering problem in the atmospheric sciences using a dragon: you might see a dragon and try to predict, how its footprints look like. But seeing just the footprints, can you anymore identify that they belonged to a dragon? Trying to predict the footprints of a dragon is known as the forward problem. When all we have is the footprints and we should guess who produced them, it is called a backward problem. Similarly, in atmospheric sciences, knowing how atmospheric particles scatter light is the forward problem. Retrieving the particle properties from their light scattering signal is the backward problem.

The backward problem is commonplace in many fields where light scattering is used as a method to measure or retrieve particle properties. These methods include satellite (e.g. Mishchenko and Travis, 1997; Baum et al., 2000; Platnick et al., 2003) and lidar observations (e.g. Sassen, 1991; Wandinger et al., 2002; Ansmann et al., 2011; Shimizu et al., 2004) as well as single particle measurements that use light scattering to detect cloud (e.g. Wendisch and Brenguier, 2013) or aerosol particles (e.g. Schwarz et al., 2006; Cai et al., 2008; Hinds, 2012). In each of these backward problems, the atmospheric particles are idealised with model particles; like a dragon could be portrayed with a model dragon. The easiest model is a sphere (see e.g. Plass and Kattawar, 1971), as the problem has an analytical solution. However, we cannot assume a dragon to really be spherical; similarly, we cannot assume all the atmospheric particles to be spheres. It has long been known that assuming a spherical shape could lead to significant errors in modelling the climate change by tropospheric particles (Kahnert et al., 2005). The importance of appropriate and accurate particle models cannot be stressed enough, as these models will be used in the radiative transfer calculations. For example, just the difference in ice crystal habits could cause the radiative forcing of cirrus with moderate optical thickness to vary over 26% (Wendisch et al., 2005).

In the field of atmospheric ice particles, the development of model particles has long oriented towards accurate representing of the ice crystal shapes. In-situ observations have indicated that ice clouds are typically composed of hexagonal plates, columns and bullet rosettes (e.g. Heymsfield and Platt, 1984). Therefore, a simple hexagonal model was introduced to calculate the scattering properties of non-spherical ice crystals in the 1970s and 1980s (Wendling et al., 1979; Cai and Liou, 1982; Takano and Liou, 1989). As the computational power increased, also other habit models started to emerge (e.g. Macke, 1993; Takano and Liou, 1995; Nousiainen and McFarquhar, 2004; Um and McFarquhar, 2009). However, introducing different shapes was not the end of the development of particle models.

After polarisation resolved satellite measurements were introduced, it became clear that atmospheric ice crystals do not necessarily have simple pristine shapes (Diedenhoven et al., 2012; Cole et al., 2014). The fraction of the backward scattered light was typically higher and the angular degree of polarisation smoother than any pristine particle model would have predicted. This brought the modellers back to the drawing board and the next generation of particle models explained these observations by inducing structural complexity by aggregation or surface modulation to the model particles (e.g. Um and McFarquhar, 2007; Yang et al., 2008; Baum et al., 2010). Soon, it was suggested that it might not be necessary to have a large selection of different particle shapes, but only a few habits could predict the observations, if complexity was included (Liu et al., 2014).

On one hand, it seemed that the particle shape might not be so important after all in determining the radiative and scattering properties of the ice particles. On the other hand, the new particle models have been criticised using complexity as an arbitrary tuning parameter that might not have any relation to real atmospheric ice particles. Therefore, the atmospheric scientists are still left with two open questions: is there such a thing as ice particle complexity and what does really determine the light scattering properties of atmospheric particles: is it their shape, their degree of complexity, or both?

It is not just ice particles, whose scattering properties are sensitive to their morphology. Complex aerosol particles are normally approximated with simple geometrical forms (e.g. Dubovik et al., 2006), although severe discrepancies in light scattering properties exist between the models and real atmospheric particles (Merikallio et al., 2011). Similar to ice particles, assuming a non-spherical shape for the aerosol particles significantly increases the intensity of the light scattered into the backward hemisphere (Mishchenko et al., 1995a).

Whether it is evaluating the performance of a particle model or a radiative transfer model, or trying to find a solution to a backward problem, these all rely on solving the forward problem first. This is where experimental particle optics come into the picture. Especially laboratory experiments can simplify the forward problem; we can remove the dragon from its surroundings. However, also experimental work in the field of atmospheric optics is facing challenges. How to study and define the complexity of atmospheric particles? Non-volatile aerosol particles can easily be removed from the surrounding gas and studied under the electron microscope but how do you capture an ice crystal to look at its fine structure? Even if new measurement techniques are emerging, it is not enough to apply them only in one location. We need a global picture.

1.1. Summary of original contributions

This thesis focuses on the angular light scattering properties of complex atmospheric particles. We take an experimental approach to try to solve the aforementioned forward problem. We limit us to single-scattering investigations at visible wavelengths (short-wave scattering). I present series of experiments, where the angular light scattering properties of aerosol and ice particles are studied. Below, I outline the major contributions.

Measurements of the near-backscattering depolarisation ratio. My first contribution to understand the light scattering properties of various atmospheric particles was to build an in-situ scattering and depolarisation instrument. Depolarisation measurements can be used to detect aspherical particles and to distinguish between different types of atmospheric particles. I deployed this instrument in laboratory experiments, where I measured the linear and circular depolarisation ratio of dust particles, secondary organic aerosol particles and ice particles of various origins.

Angular light scattering of complex ice particles. I was involved in the last phases of the development of the novel Particle Habit Imaging and Polar Scattering (PHIPS) instrument and operated it during the first field and laboratory campaigns. With this instrument, it is possible to measure the angular scattering of atmospheric ice particles, part of the so-called scattering phase function. I investigated the influence of ice particle complexity on this optical quantity.

Field observations of ice particle complexity. A newly developed in-situ method can distinguish ice crystal small-scale complexity, like surface roughness. The method was tested in laboratory experiments, and in this thesis I applied the method in field measurements at three climatically distinct regions: in the arctic, the mid-latitude regions, and the tropics.

1.2. Outline of this thesis

This thesis is divided into seven chapters:

Chapter 2 gives the theoretical framework of this work. It begins with the fundamentals of light scattering theory of small particles. Afterwards, the applications of light scattering theory in the context of atmospheric particles are discussed.

Chapter 3 presents the experimental methods used in this thesis and describes the measurement platforms.

In **Chapter 4** I show laboratory measurements of the depolarisation properties of mineral dust aerosol and viscous secondary organic particles.

In **Chapter 5** I move from complex aerosol particles to study the light scattering by complex ice particles. This chapter focuses on laboratory studies of hexagonal and near-spherical ice particles.

In **Chapter 6** the knowledge gained from the laboratory studies will be applied to field measurements of the morphology and scattering properties of ice crystal with different origins.

Chapter 7 discusses the results gained from chapters 4, 5 and 6 and the possible outlooks and implications of these results for future climate predictions.

2. Background

2.1. Light Scattering by Small Particles

As a parallel monochromatic beam of light propagates in a vacuum it does not change its intensity or polarisation state. However, would the beam encounter a particle, several interactions can be expected. First, the particle could convert part of the beams energy into other energy forms, like heat; this process is called *absorption*. Second, the particle could scatter some of the beam energy to all directions with respect to the beam propagation without changing the frequency; this process is called *elastic scattering*. When the frequency is changing in the scattering process the scattering is *inelastic* or non-linear. Inelastic or non-linear scattering processes, however, will not be considered in this thesis. Absorption and scattering reduce the beam energy and this reduction in the energy is called *extinction*. Different polarisation components may have different extinction rates, which could lead to the change in the polarisation state of the beam after it have encountered the particle.

In electromagnetic terms, the light beam can be considered as an oscillating electromagnetic wave and the particle as a large collection of discrete electric charges: electrons and protons. The oscillating electromagnetic field excites the charges in the particle to oscillate in the same frequency. The oscillating charges emit secondary electromagnetic waves and the superposition of these secondary waves gives the elastically scattered field. As the number of charges in a particle is enormous, it is not meaningful to calculate the scattered field by superposing the microscopic secondary waves. Therefore, for scattering problems macroscopic electromagnetics are used that consider the large collection of microscopic charges as a macroscopic body. The scattering problem can be solved using the Maxwell equations to describe the macroscopic field that is subject to a body with appropriate boundary conditions.

In general, light scattering can be divided in three domains with respect to the dimensionless *size parameter*, x, which is defined as

$$x = \frac{\pi d_p}{\lambda}, \tag{2.1}$$

where d_p is the particle diameter and λ the wavelength of the light. The three domains are then

$x \gg 1$: Rayleigh scattering
$x \approx 1$: Mie scattering
$x \ll 1$: Geometric scattering

Fig. 2.1 shows the different scattering regimes with respect to the light wavelength and the particle size. Rayleigh scattering is elastic scattering that describes the scattering from particles that are much smaller than the wavelength of the light, e.g. scattering from atoms or molecules. For example the the Rayleigh scattering of atmospheric molecules is responsible for the blue colour of the sky. Mie scattering is scattering of particles that have sizes roughly corresponding to the wavelength. In the visible wavelengths, the Mie scattering covers accumulation and coarse-mode aerosol particles, cloud droplets and ice particles up to 1 mm. Rayleigh particles have a scattering phase function that is symmetric with respect to the fraction of forward and backward scattered light. In Mie scattering the fraction of forward scattered light intensity increases with increasing particle dimensions. Therefore, in the forward angular range the Mie scattering can dominate the Rayleigh scattering. Also, the larger the body the more it scatters, i.e. the direct radiative effect of atmospheric constituents is largely determined by the particle scattering in the Mie regime.

The Maxwell equations can be solved analytically in the case of spherical boundary conditions. This solution was found by Gustav Mie (Mie, 1908) and Ludvig Lorenz and later further developed by many others. The theory is named after its developers and is called the Lorenz-Mie theory, although also names like Mie-theory or Lorenz-Mie-Debye theory are used. Analytical solution can be also found for other simple geometries, like for infinite long cylinders, but for complex particles the Maxwell equations need to be solved numerically. One of the most well-known methods for computing the electromagnetic scattering by complex particles is the T-matrix approach, also known as extended boundary condition method or null-field method (Waterman, 1965; Mishchenko et al., 1996). Other numerical methods include the finite-difference time-domain method (Yee et al., 1966; Yang and Liou, 2000), the finite-element method (Morgan and Mei, 1979), the method of lines (Pregla, 1989), the point matching method (Oguchi and Hosoya, 1974) and the discrete dipole approximation (DDA) (DeVoe, 1964, 1965; Purcell and Pennypacker, 1973). *Multiple scattering* complicates the scattering process further and, although frequent in the atmosphere (e.g. multiple scattering in clouds), most of the numerical solutions assume only *single scattering*, which means that each photon is scattered only once.

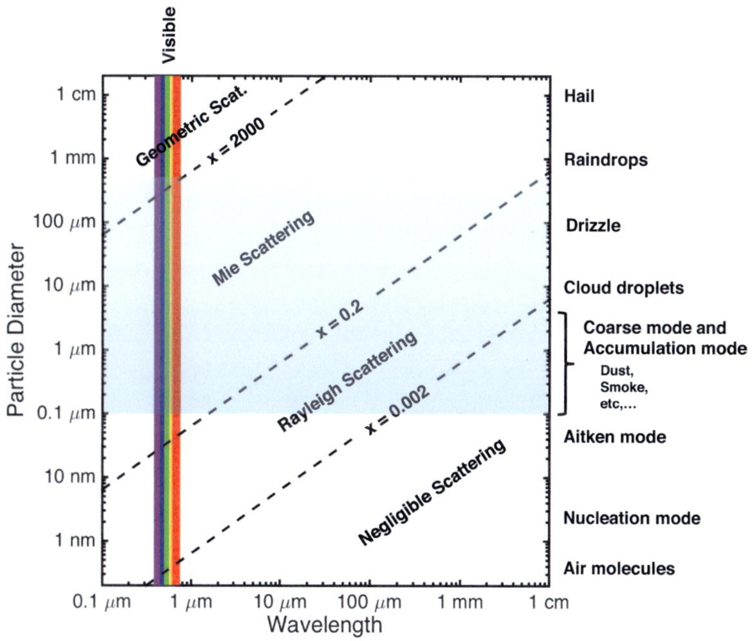

Figure 2.1.: Illustration of the different scattering regimes. The blue shaded area shows the particle size range that this thesis addresses.

2.1.1. The Maxwell equations and scattering problem in the case of a plane wave

The macroscopic electromagnetic field can be described using the Maxwell equations, which in SI units may be written as

$$\nabla \cdot \vec{E} = \frac{\rho}{\epsilon_0} \quad \text{(Gauss's law)} \tag{2.2}$$

$$\nabla \cdot \vec{B} = 0 \quad \text{(Gauss's law for magnetism)} \tag{2.3}$$

$$\nabla \times \vec{E} = -\frac{\partial B}{\partial t} \quad \text{(Faraday's law of induction)} \tag{2.4}$$

$$\nabla \times \vec{B} = \mu_0 \vec{J} + \mu_0 \epsilon_0 \frac{\partial E}{\partial t} \quad \text{(Ampére's circuital law)} \tag{2.5}$$

where $\vec{E}$ is the electric field, $\vec{B}$ the magnetic field, ρ the electric charge density, ϵ_0 the permittivity of free space, $\vec{J}$ the electric current density and μ_0 the permeability of free space.

The Maxwell equations can be simplified in the case of a plane wave. The definition for plane wave is a monochromatic parallel beam of light and it is an idealised description of wave propagation. However, it can be used as an approximation at *far-field* region, which covers most of the light scattering applications. The solution for a plane wave can be given as following

$$\vec{E}_c(\vec{r}, t) = \vec{E}_0 e^{i\vec{k}\cdot\vec{r} - i\omega t} \qquad \vec{H}_c(\vec{r}, t) = \vec{H}_0 e^{i\vec{k}\cdot\vec{r} - i\omega t} \tag{2.6}$$

where $\vec{E}_0$ and $\vec{B}_0$ are constant complex vectors, $\vec{k}$ a constant wave vector, $\vec{r}$ the position vector and ω the angular frequency. The Maxwell equations for a plane wave now take the following form

$$\vec{k} \cdot \vec{E}_0 = 0 \tag{2.7}$$

$$\vec{k} \cdot \vec{B}_0 = 0 \tag{2.8}$$

$$\vec{k} \times \vec{E}_0 = \omega \mu \vec{B}_0 \tag{2.9}$$

$$\vec{k} \times \vec{B}_0 = -\omega \epsilon_0 \vec{E}_0 \tag{2.10}$$

The first two equations tell that both $\vec{E}_0$ and $\vec{B}_0$ are perpendicular to $\vec{k}$, i.e. the traveling direction of the electromagnetic wave. The equations 2.9 and 2.10 show that $\vec{E}_0$ and $\vec{B}_0$ are also mutually perpendicular: $\vec{E}_0 \cdot \vec{B}_0 = 0$.

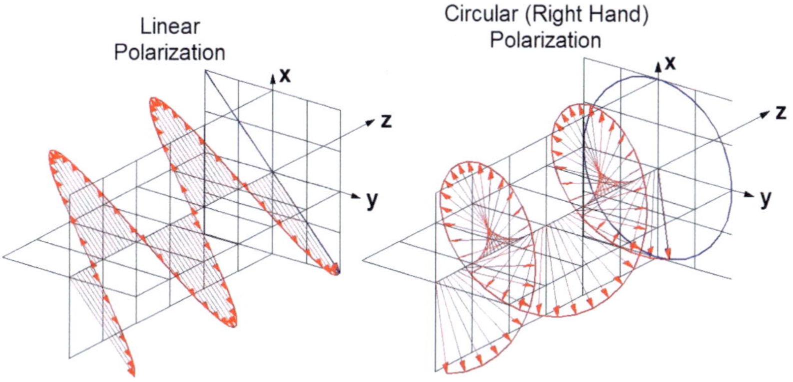

Figure 2.2.: Linearly and circularly polarised light. By Mehmet Yavuz.

2.1.2. Polarised light and Stokes parameters

Light can be considered as a plane wave with both the electric and magnetic fields oscillating perpendicularly to the traveling direction. The oscillation direction of the electric field determines the *polarisation* of the light. Normally, natural light is composed of many electromagnetic waves, whose electric fields have various oscillation directions with the equal probabilities. Such light is called *unpolarised*. If the electric field is oscillating in only one direction, the light is called *linearly polarised*. If the electric field is rotating at a frequency but keeps its amplitude constant, then the light is said to be *circularly polarised*.

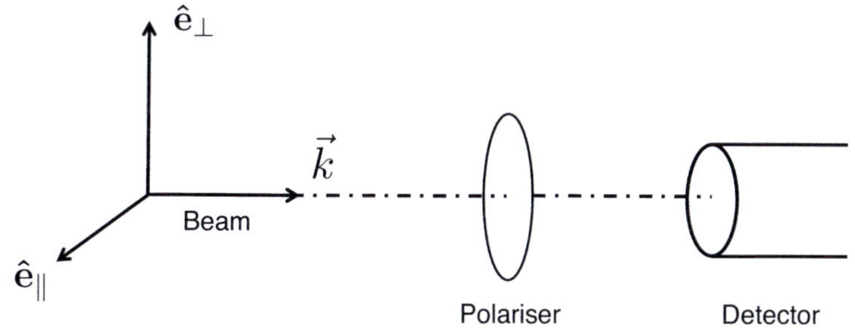

Figure 2.3.: Detector measuring the beam transmitted by the polariser.

The polarisation state of the light can be described using the *Stokes parameters*. They were defined by George Gabriel Stokes in 1852 to mathematically describe polarisation in the terms of the total intensity (I), degree of linear polarisation (Q and U) and degree of circular polarisation (V). The Stokes parameters are particularly convenient because they all are measurable quantities. Here, I derive the Stokes parameters in a linear basis by considering a simple experiment with a detector and a polariser (Fig. 2.3). The incoming electric field can be considered to have horizontal and vertical components, so that

$$\vec{E_0} = E_{\parallel}\hat{\mathbf{e}}_{\parallel} + E_{\perp}\hat{\mathbf{e}}_{\perp} \tag{2.11}$$

where $\hat{\mathbf{e}}_{\parallel}$ and $\hat{\mathbf{e}}_{\perp}$ are unit vectors for the horizontal and vertical directions. We can perform four experiment with different polarisers:

I Intensity measurement without polariser. The first Stokes parameter I describes the total intensity of the wave. The total intensity I can be related to the electric field strength E as $I = E^*E$, where E^* is complex conjugate of E

$$I = E_{\parallel}E_{\parallel}^* + E_{\perp}E_{\perp}^*. \tag{2.12}$$

II Horizontal and vertical linear polarisers. If the polariser is horizontally aligned, we measure the intensity $E_{\parallel}E_{\parallel}^*$ and if the polariser is vertically aligned, the intensity will be $E_{\perp}E_{\perp}^*$. The difference between these two measurements is then

$$Q = I_{\parallel} - I_{\perp} = E_{\parallel}E_{\parallel}^* - E_{\perp}E_{\perp}^*. \tag{2.13}$$

III +45° and -45° linear polarisers. If the polariser is aligned ±45° from the $\parallel$-direction, the $\parallel$- and $\perp$-components of the electric field are equally transmitted, i.e.

$$E_+ = \frac{1}{\sqrt{2}}(E_{\parallel} + E_{\perp}) \qquad E_- = \frac{1}{\sqrt{2}}(E_{\parallel} - E_{\perp}) \tag{2.14}$$

The resulting intensities are

$$E_+E_+^* = \frac{1}{2}(E_{\parallel}E_{\parallel}^* + E_{\parallel}E_{\perp}^* + E_{\perp}E_{\parallel}^* + E_{\perp}E_{\perp}^*) \tag{2.15}$$

and

$$E_-E_-^* = \frac{1}{2}(E_{\parallel}E_{\parallel}^* - E_{\parallel}E_{\perp}^* - E_{\perp}E_{\parallel}^* + E_{\perp}E_{\perp}^*). \tag{2.16}$$

The third Stokes parameter U is again the difference of these two signals.

$$U = I_+ - I_- = E_{\parallel}E_{\perp}^* + E_{\perp}E_{\parallel}^* \tag{2.17}$$

IV Circular polarisers. In the case of circular polarisation, the light is propagating with two equally high amplitudes $E_{\|}$ and $E_{\perp}$, but those two amplitudes have a $\pi/2$ phase shift. Therefore, I introduce a new set of basis vectors $\hat{\mathbf{e}}_R$ and $\hat{\mathbf{e}}_L$. The incident field can now be written as $\vec{E}_0 = E_R\hat{\mathbf{e}}_R + E_L\hat{\mathbf{e}}_L$, where

$$E_R = \frac{1}{\sqrt{2}}(E_{\|} - iE_{\perp}) \qquad E_L = \frac{1}{\sqrt{2}}(E_{\|} + iE_{\perp}). \tag{2.18}$$

Now, if the polariser is right-handed polarised the transmitted intensity I_R is

$$I_R = E_R E_R^* = \frac{1}{2}(E_{\|}E_{\|}^* + iE_{\|}E_{\perp}^* - iE_{\perp}E_{\|}^* + E_{\perp}E_{\perp}^*) \tag{2.19}$$

and if the polariser is left-handed polarised the transmitted intensity I_L is

$$I_L = E_L E_L^* = \frac{1}{2}(E_{\|}E_{\|}^* - iE_{\|}E_{\perp}^* + iE_{\perp}E_{\|}^* + E_{\perp}E_{\perp}^*). \tag{2.20}$$

The last Stokes parameter V is the difference

$$V = I_R - I_L = i(E_{\|}E_{\perp}^* - E_{\perp}E_{\|}^*). \tag{2.21}$$

The Stokes parameters are often combined into a vector, known as the *Stokes vector*:

$$\vec{I} = \begin{pmatrix} I \\ Q \\ U \\ V \end{pmatrix} = \begin{pmatrix} E_{\|}E_{\|}^* + E_{\perp}E_{\perp}^* \\ E_{\|}E_{\|}^* - E_{\perp}E_{\perp}^* \\ E_{\|}E_{\perp}^* + E_{\perp}E_{\|}^* \\ i(E_{\|}E_{\perp}^* - E_{\perp}E_{\|}^*) \end{pmatrix} \tag{2.22}$$

With the Stokes vector it is possible to describe the polarisation state of the light beam. For horizontally polarised light $Q > 0$ is valid and for vertically polarised light $Q < 0$. For +45° polarised light $U > 0$ is valid and for -45° polarised light $U < 0$. Right-handed circular circular polarisation fulfills $V > 0$ and left-handed $V < 0$. For ideally polarised light the Stokes vectors can be described as in the table 2.1.

From the Stokes parameters we can see that for polarised light it is valid that

$$I^2 = Q^2 + U^2 + V^2. \tag{2.23}$$

Lin. polarised (horizontal)	Lin. polarised (vertical)	Circ. polarised (right-handed)	Circ. polarised (left-handed)	Unpo- larised
$\begin{pmatrix}1\\1\\0\\0\end{pmatrix}$	$\begin{pmatrix}1\\-1\\0\\0\end{pmatrix}$	$\begin{pmatrix}1\\0\\0\\1\end{pmatrix}$	$\begin{pmatrix}1\\0\\0\\-1\end{pmatrix}$	$\begin{pmatrix}1\\0\\0\\0\end{pmatrix}$

Table 2.1.: Stokes vectors for different polarisation states.

We can also determine the degree of linear polarisation as

$$P_L = -\frac{Q^2 + U^2}{I^2} \tag{2.24}$$

and the degree of circular polarisation as

$$P_C = \frac{V}{I}. \tag{2.25}$$

2.1.3. The scattering matrix

The Stokes vector of the scattered light can be related to the Stokes vector of the incident light by means of a linear transformation

$$\begin{pmatrix} I_{sca} \\ Q_{sca} \\ U_{sca} \\ V_{sca} \end{pmatrix} = \frac{1}{k^2 r^2} \mathbf{S} \begin{pmatrix} I_{inc} \\ Q_{inc} \\ U_{inc} \\ V_{inc} \end{pmatrix} \tag{2.26}$$

where k is the wave number, r the distance to the detector and **S** is a 4×4 *scattering matrix* having 16 real independent elements. For a general case the scattering matrix can be given as

$$\begin{pmatrix} S_{11}(\theta,\phi) & S_{12}(\theta,\phi) & S_{13}(\theta,\phi) & S_{14}(\theta,\phi) \\ S_{21}(\theta,\phi) & S_{22}(\theta,\phi) & S_{23}(\theta,\phi) & S_{24}(\theta,\phi) \\ S_{31}(\theta,\phi) & S_{32}(\theta,\phi) & S_{33}(\theta,\phi) & S_{34}(\theta,\phi) \\ S_{41}(\theta,\phi) & S_{42}(\theta,\phi) & S_{43}(\theta,\phi) & S_{44}(\theta,\phi) \end{pmatrix} \tag{2.27}$$

where the elements $S_{m,n}$ are intensive parameters which depend on the size parameter, particle shape, refractive index, the polar scattering angle θ and the azimuthal scattering angle ϕ. When randomly oriented particles and their mirror particles are present in equal numbers in the ensemble the scattering matrix has a simpler form (Hovenier et al., 2003)

$$\begin{pmatrix} S_{11}(\theta) & S_{12}(\theta) & 0 & 0 \\ S_{12}(\theta) & S_{22}(\theta) & 0 & 0 \\ 0 & 0 & S_{33}(\theta) & S_{34}(\theta) \\ 0 & 0 & -S_{34}(\theta) & S_{44}(\theta) \end{pmatrix} \tag{2.28}$$

The matrix element S_{11} represents the total intensity of the scattered light at a certain angle θ

$$I = \frac{1}{k^2 r^2} S_{11}(\theta) I_0. \tag{2.29}$$

and it can be linked to the *differential scattering cross section* through

$$\frac{dC(\theta)}{d\Omega} = \frac{S_{11}}{k^2}. \tag{2.30}$$

The differential scattering cross section specifies the angular distribution of the scattered light, i.e. the amount of light scattered into a unit solid angle of a given direction (Bohren and Huffman, 2008). In light scattering theory, one commonly encounters the term *phase function*. The phase function can be related to the differential scattering cross section as following

$$p = \frac{1}{C_{sca}} \frac{dC(\theta)}{d\Omega}, \tag{2.31}$$

where C_{sca} is the total scattering cross section. The phase function, or scattering phase function, gives the angular distribution of light intensity scattered by a particle at a given wavelength. From Eq. 2.31 it is obvious that the integral of p over the complete hemisphere results in unity:

$$\int_{4\pi} p \, d\Omega = 1. \tag{2.32}$$

The asymmetry of the scattering phase function (light scattered to the forward direction to light scattered to the backward direction) is defined with the *asymmetry parameter*, g, that is the average cosine of the scattering angle

$$g = \int_{4\pi} p\, cos(\theta) d\Omega. \tag{2.33}$$

Figure 2.4.: Electron microscope images and images taken by a cloud particle imager (CPI) from atmospheric particles.

2.2. Atmospheric particle shape

Our atmosphere is full of solid, semisolid or water soluble bodies that take up myriad of shapes. The variability in the particle size and shape spectrum is depicted in Fig. 2.4. The simplest atmospheric particles take the form of a sphere as in the case of water soluble particles (Ebert et al., 2002) or pure water droplets. Other typical forms are faceted particles, as in the case of ice crystals, or particles with a completely randomised form, if they are formed as aggregates (e.g. soot) or if they are fractal particles (e.g. dust) (Nousiainen, 2009; Bingemer et al., 2012). In this thesis we concentrate on a few atmospherically relevant particle types that all have a non-spherical or complex shape: dust particles, secondary organic aerosol particles and ice particles.

Irregular mineral dust particles are ubiquitous in the Earth's atmosphere (e.g. Sokolik and Toon, 1996; Andreae and Merlet, 2001; Prospero et al., 2002). The size distribution of atmospheric dust particles covers a wide range from a few hundred nanometers to several microns coarse-mode particles (Weinzierl et al., 2009), which makes these particles large enough that the can efficiently scatter and absorb sun light. Therefore, it is important to asses the radiative impact of dust correctly. For this, the information of the dust particle single-scattering properties as well as their global coverage are needed. The global coverage of dust is retrieved using remote sensing methods that, also, rely on the accurate description of the mineral dust particle radiative properties. To accurately represent the dust scattering properties both in climate models and in remote sensing retrievals, we need to understand how these irregular particles interact with light. For this purpose, laboratory studies on the optical properties of mineral dust particles have proven to be useful (e.g. Linke et al., 2006; Muñoz et al., 2007, 2010; Wagner et al., 2012). Among the optical properties that can be measured in the laboratory, the backscattering linear depolarisation ratio is of particular importance for the interpretation of polarisation lidar observations (Sakai et al., 2010).

Besides mineral dust particles, organic particulate material is abundant in Earth's atmosphere. Biogenic and anthropogenic sources emit volatile organic compounds (VOCs), which are oxidised through a cascade of chemical reactions into extremely low volatility vapours that condense into the particle phase to form secondary organic aerosol (SOA) (Hallquist et al., 2009). Biogenic VOCs are much more abundant than anthropogenic VOCs (Guenther et al., 1995; Jimenez et al., 2009) and monoterpenes such as α-pinene are found throughout the continental boundary layer, particularly in boreal forest regions (e.g., Tunved et al., 2006; Laaksonen et al., 2008). After formation, SOA influences climate on a global scale directly by scattering and absorbing solar radiation and indirectly through aerosol-cloud interactions. Locally, SOA can affect air quality and human health (e.g., Nel, 2005; Huang et al., 2014). However, the chemical and physical processes that determine the properties of SOA particles are complex, and our understanding of these processes is limited (Hallquist et al., 2009; Hoyle et al., 2011).

The largest of the atmospheric particulate constituents are water particles in clouds. Cloud particles can be found in warm clouds, where they take a liquid form, in mixed-phase clouds with both ice particles and water droplets, and in cold ice clouds, like cirrus clouds that are completely glaciated. Cold ice clouds have a large temporal and spatial distribution in the atmosphere and are known to have a significant effect on Earth's climate (e.g. Stephens et al., 1990; Nazaryan et al., 2008). The magnitude of this effect depends among other on the ice crystal morphology (e.g. shape), particle size distribution (especially the distribution of small ice particles) and the spatial distribution of ice shapes and sizes as a function of distance from the cloud top (Wendisch et al., 2005; Mitchell et al., 2008; Yang et al., 2012; Yi et al., 2013; Baran et al., 2015). Ice clouds consists of ice crystals with variety of shapes, sizes and degrees of crystal complexity (surface distortions or geometrical complexity). In-situ measurements have shown that the geometrical form of ice crystals typically deviates from spherical form and can range from symmetrical hexagonal columns, plates and bullets to non-symmetrical ice particles and to aggregates of the above mentioned forms (e.g. Korolev et al., 1999; Korolev and Isaac, 2003; Connolly et al., 2005; Heymsfield et al., 2005; Lawson et al., 2006). In addition to these in-situ observations of ice particle shapes, various indirect measurements have been done to investigate the ice crystal complexity in natural ice clouds. These include satellite observations that have indicated that ice crystal complexity is a globally important feature in cirrus clouds (Baran et al., 2015; Cole et al., 2014), whereas laboratory experiments have shown that smaller scale complexities, such as surface roughness, is present in ice crystals under a wide range of environmental conditions (Baran et al., 2011; Magee et al., 2014). However, direct in-situ observations of ice particle complexity are extremely rare and limited to case studies (Ulanowski et al., 2014).

2.3. Detection of particle non-sphericity

The polarisation lidar technique is commonly used to study the vertical profiles of both tropospheric and stratospheric clouds (Sassen, 1991) and aerosol distributions (e.g. Ansmann et al., 2005, 2012). A key measurement quantity in this technique is the particle backscattering linear depolarisation ratio, δ_L, that describes the ratio of cross-polarised to polarised backscattering intensity. For a spherical isotropic particle, this ratio is zero (Mie, 1908), but for a irregular particle the ratio depends on the size, shape and refractive index of the particle (Mishchenko et al., 1996). Therefore, the depolarisation ratio is considered to be an indicator of the particle non-sphericity, and is used to distinguish between liquid, mixed-phase and ice clouds as well as between aerosol types.

2.3.1. Definition of the depolarisation ratio

For a spherical particle the ratio S_{22}/S_{11} is always 1 and, therefore, the departure of S_{22}/S_{11} from unity can be considered as a sign of nonsphericity and the difference is referred as the *degree of (linear) depolarisation*

$$\Delta_L = 1 - \frac{S_{22}}{S_{11}}. \tag{2.34}$$

In the lidar community, a more common measure for the deviation from the spherical form is the *linear depolarisation ratio* that at the exact backscattering angle (180°) can be defined as (see derivation in Appendix B)

$$\delta_L = \frac{S_{11} - S_{22}}{S_{11} + S_{22}}. \tag{2.35}$$

The linear depolarisation ratio can be related to the degree of linear depolarisation through

$$\delta_L = \frac{\Delta_L}{2 - \Delta_L}. \tag{2.36}$$

Similarly, we can define the *circular depolarisation ratio* at the exact backscattering angle and with right-handed circular incident polarisation as

$$\delta_C = \frac{S_{11} + S_{44}}{S_{11} - S_{44}}. \tag{2.37}$$

The linear depolarisation ratio can vary between 0 and 1, whereas the circular depolarisation ratio can reach arbitrarily large values. Mishchenko and Hovenier (1995b) showed that for mirror symmetrical randomly oriented particles the circular depolarisation ratio can be linked to the linear depolarisation ratio through the equation

$$\delta_C = 2\delta_L/(1 - \delta_L). \tag{2.38}$$

This equation tells us that the circular depolarisation is more sensitive to changes in the depolarisation properties than the linear depolarisation ratio. Therefore, some lidar applications have taken advantage of this and are measuring with circular incident polarisation (e.g. Del Guasta et al., 2006; Hayman et al., 2012).

2.3.2. Depolarisation properties of atmospheric particles

The depolarisation properties of atmospheric particles depend on their size, shape and refractive index, although it has to be noted that no linear or explicit dependency exist between the particle properties and their depolarisation ratio (Mishchenko et al., 1996). If the particles are non-spherical they will induce a non-zero depolarisation ratio. In general, atmospheric particles can be divided to those that are weakly depolarising, showing maximum depolarisation ratios of only few percent, and those that are strongly depolarising with depolarisation ratios of several tens of percent. Examples of low depolarising aerosol particles are soot aerosol (Minutolo et al., 1994; Järvinen et al., 2016a), whereas relatively high depolarisation values are measured for dust particles and for ice particles (e.g. Freudenthaler et al., 2009; Sassen, 1991; Ansmann et al., 2012).

In the SAMUM field experiments in northern Africa (Heintzenberg, 2009; Ansmann et al., 2011b) the knowledge of the dust depolarisation properties was advanced significantly. It was found that at the wavelength of 532 nm atmospheric dust particles have a relatively stable depolarisation around 0.3 (Esselborn et al., 2009; Freudenthaler et al., 2009; Groß et al., 2011). Similar depolarisation values have also been measured for dust particles in other locations (e.g. Sakai et al., 2000; Sugimoto et al., 2003; Shimizu et al., 2004; Sugimoto and Lee, 2006) and in laboratory experiments (Järvinen et al., 2016a; Miffre et al., 2016).

For atmospheric ice clouds the typical depolarisation ratios range between 0.3 and 0.4 (Schotland et al., 1971; Noel et al., 2004; Martins et al., 2011; Schnaiter et al., 2012), although also extremely low values of few percentage (Del Guasta and Niranjan, 2001) or extremely high values above 0.5 have been reported in the case of contrail cirrus (Freudenthaler et al., 1995). The link between ice crystal size, shape and their depolarisation ratio relies on the observations and modelling studies. For particles that are smaller than few tens of micrometer, their depolarisation properties can be calculated using the T-matrix method (Mishchenko et al., 1996). With this method it was shown that the high depolarisation values can be explained with small, micrometer-sized ice particles (Mishchenko and Sassen, 1998). However, the link between the particle properties and its depolarisation ratio is still poorly understood. Open questions remain, like how does the depolarisation properties of pristine ice crystals differ from those that are complex.

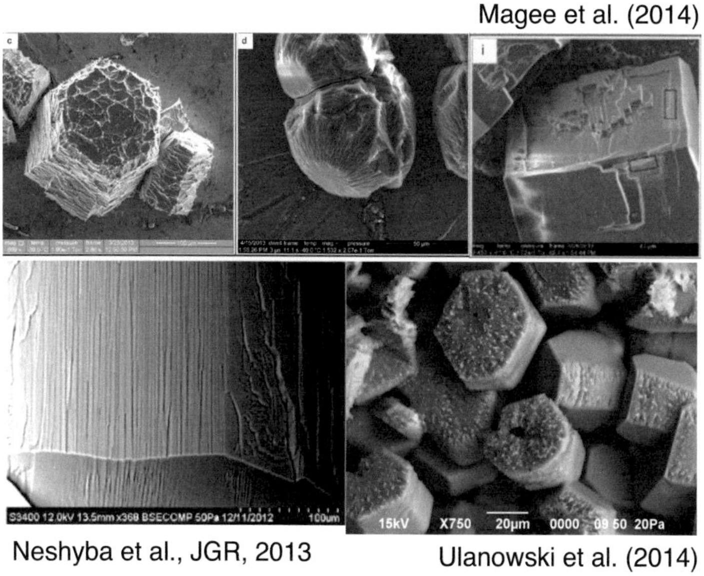

Figure 2.5.: Examples of ice particle surface microstructures.

2.4. What is ice particle complexity?

A significant amount of research interest have been focused to study the geometrical form of the ice particles in different environmental conditions or measurement locations (e.g. McFarquhar and Heymsfield, 1996; Heymsfield and Aulenbach, 1998; Lawson et al., 2006; Bailey and Hallett, 2009; Baran, 2009) but due to the limitations of the ice particle imaging methods, this analysis has been limited to larger scale features like polycrystal or dentritic structures. However, indications of the importance of small scale ice particle structures have been long available. Several studies have implemented interference techniques to suggest micro scale distortions in the ice facets (Bryant et al., 1960; Sazaki et al., 2010) and growing amount of cirrus radiative studies have concluded that a better agreement between models and observations is established if including complexity to the particle models (e.g. Baum et al., 2011; Mauno et al., 2011; Liu et al., 2014).

After introducing scanning electron microscopy (SEM) imaging techniques to the atmospheric sciences, the micro scale structure of ice particle could be imaged directly. SEM studies, like those by Neshyba et al. (2013); Magee et al.

(2014); Ulanowski et al. (2014), have shown that small-scale surface structures are common in the ice particles under all conditions (Fig. 2.5). However, the limitation of the these studies is that it is difficult to apply for atmospheric ice particles and, furthermore, the environmental conditions in the SEM differ from those found in the atmosphere.

Micro-structures add to the overall complexity of single ice particle habits. In the remaining of this thesis I refer this kind of complexity as the *small-scale complexity*. Complexity can also be induced on larger scale by forming aggregated particles. Such processes are common in mixed-phase environments, where the ice particle growth is rapid and particle concentrations are high, which lead to collisions between the cloud particles. One of the processes taking place in mixed-phase clouds is riming, where supercooled liquid droplets collide with existing ice particles and freeze in the process. This leads to an aggregate of the original ice particle and a frozen droplet with usually much smaller dimensions. Riming is an efficient way to grow the ice particle mass, which eventually leads to graupel particles that later precipitate.

The mixed-phase environment also promotes the formation of smaller ice particle aggregates, like chain-aggregates, that in convective systems are transported to anvil cirrus. Such particles are frequently measured in tropical convective systems (e.g. Connolly et al., 2005; Frey et al., 2011), where they can be imaged with traditional cloud particle imagers. Therefore, in contrast to small-scale complexity, in the remainder of this thesis I label ice particle aggregation as the *large-scale complexity*.

2.4.1. Quantifying the degree of ice particle complexity

Investigations of ice particle complexity is often based on analysis of particle images taken with a cloud particle imager (CPI). The resolution of such images is restricted to few micrometers, which only makes investigation of large-scale complexity possible. The simplest measure of particle complexity can be considered to be the deviation from the spherical shape. From the CPI measurements it is possible to determine the particle asphericity using the area ratio α (i.e., the projected area of a particle divided by a circumscribed circle with diameter D_{max}) (McFarquhar et al., 2013). With a suitable cutoff α it is possible to discriminate between spherical and aspherical particles. Schmitt and Heymsfield (2014) further developed analysis methods for the CPI images and introduced a dimensionless complexity parameter, C, for representing the crystal complexity (see details in Sec. 3.3.3). The authors showed that with the complexity parameter C it was possible to discriminate between single ice particle habits and aggregated ice particles.

Quantifying small-scale complexity from atmospheric measurements was first possible after introduction of a new measurement method that is based on recording light scattering patterns from single ice crystals. The method deploys same principle what is used in industrial quality control applications for roughened surfaces. The surface, or in this case a three dimensional particle, is illuminated with a laser beam and the spatial distribution of the scattered light is measured. The scattering pattern can be analysed with a grey level co-occurrence matrix method (GLCM) described in Lu et al. (2006). Lu et al. (2006) showed that this method is sensitive to surface features that are in the order of the used laser light wavelength. Ulanowski et al. (2010) was the first to use the method to investigate ice crystal small-scale complexity. The authors quantified the small-scale complexity using an energy parameter and a combined roughness parameters und described the relative changes in these parameters in natural ice clouds. Schnaiter et al. (2016) used the method to determine the small-scale complexity through a complexity parameter k_e (see details in Sec. 3.2.3), which was found to be more robust for the changes in small-scale complexity. The authors calibrated the complexity parameter k_e in laboratory experiments and linked the magnitude of k_e to the ice crystal growth conditions (see details in Sec. 5.1).

In modelling studies complexity of a model particle is usually induced by modulating the particle's surface (e.g. Yang and Liou, 1998; Yang et al., 2008). This is done by dividing an ice particle facet into N segments that are given different slopes specified by the first-order Gram-Charlier or the 2-D Gaussian distribution. Now, the degree of the surface roughness is determined through the σ parameter that describes the standard deviation of the distribution function. The larger the σ parameter, the wider the distribution and the higher the degree of surface roughness. Typical σ values for pristine particles with relative smooth surfaces are below 0.1 depending on the particle size and underlying shape, whereas more roughened ice particles typically have σ values above 0.1. Although the above discussed method is well established, it has been criticised for not representing the surface roughness in a realistic way.

There is no single established measure for the ice particle complexity, which makes comparison of the measurements and models difficult. The challenge is that the measured parameters, like the k_e value and the C parameter are usually instrument dependent, whereas the σ parameter might not be physically meaningful. In this work the small-scale complexity is quantified through the k_e parameter but efforts are also made to relate this measured parameter to a modelled σ value.

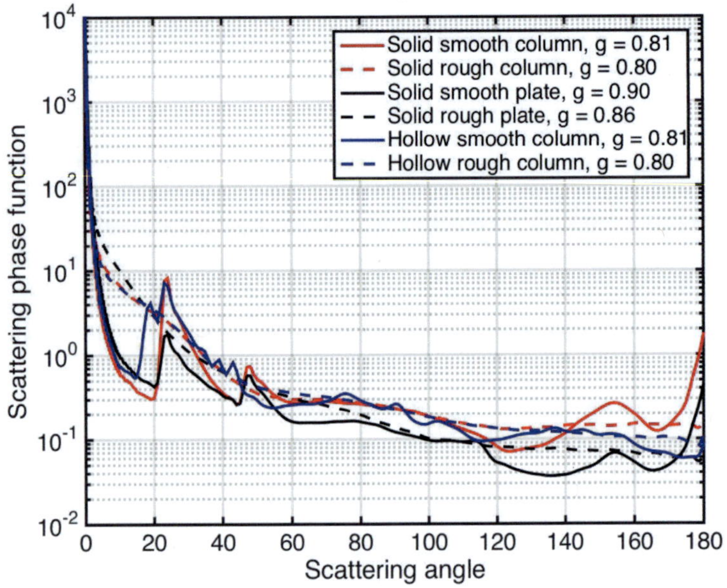

Figure 2.6.: Scattering phase functions for orientation averaged ice particle habits from Prof. P. Yang's database. All the particles have the same maximum dimensions of 100 µm.

2.4.2. Angular light scattering by complex ice particles

Both the shape of the ice crystals as well as their complexity changes the radiative properties of the ice clouds. Modelling studies (e.g. Yang et al., 2008; Baum et al., 2010) have shown that inducing crystal complexity leads to disappearance of hexagonal shape-dependent features, like 22° and 46° halo, and to smoothening of the angular scattering phase function (Fig. 2.6). Moreover, the asymmetry parameter of a roughened ice particle is smaller compared to its smooth counterpart (Ulanowski et al., 2006), which leads to a higher fraction of backscattered light. This, on the other hand, can change the radiative forcing of these ice particles (Yi et al., 2013) and leads to uncertainties in climate predictions.

Atmospheric and laboratory measurements support an ice scattering phase function with a low asymmetry parameter. Cole et al. (2014) defined a median global asymmetry parameter of 0.75 using PARASOL (Polarization & Anisotropy of Reflectances for Atmospheric Sciences coupled with Observations from a Lidar) satellite data and assuming roughened ice particles. Similar, low asymmetry parameters have been measured in-situ at different measurement campaigns around the globe. Febvre et al. (2009) measured an asymmetry parameter of 0.79 that was the same for natural cirrus and anthropogenic aged contrail cirrus. Young contrails, however, showed a significantly higher asymmetry parameter of 0.83 that can be associated with pristine habits (Fig. 2.6). During the Interhemispheric Differences in Cirrus Properties From Anthropogenic Emissions (INCA) experiment cirrus clouds were targeted at southern and northern hemisphere. Gayet et al. (2004) found a median asymmetry parameter between 0.76–0.78 that was similar for both hemispheres. Furthermore, a low asymmetry parameter of 0.78 is associated with ice crystals in the arctic (Lampert et al., 2009).

Although, the laboratory and in-situ measurements as well as the satellite retrievals show indications of the ice crystal complexity and its impact to the scattering phase function, until now simultaneous in-situ measurements of the ice crystal complexity and their scattering properties have not been conducted. This maintains a gap in the knowledge of the radiative and microphysical properties of atmospheric ice particles. For example, it is not yet proven that the low asymmetry parameter observed in ice clouds is a result of ice particle complexity. In this thesis, my motivation is to clarify these open questions.

3. Experimental Methods

In this section the fundamental experimental methods are presented and described. These methods use interactions with light and material to probe the particle shape and/or the degree of particle complexity. The first part of this chapter describes the scattering and depolarisation instrument SIMONE that can be used to detect non-spherical or inhomogeneous particles in the size range of few hundred nanometers to several hundred micrometers. This accounts cloud particles as well as aerosol particles in accumulation and coarse mode range that are responsible for the majority of the interactions with shortwave radiation in the atmosphere. The second part presents a method to study the small-scale complexity of micrometer-sized cloud and aerosol particles using two dimensional diffraction patterns recorded by the Small Ice Detector mark 3 (SID-3) and the Particle Phase Discriminator mark 2, Karlsruhe Edition (PPD-2K). In the third part the Particle Habit Imaging and Polar Scattering probe (PHIPS-HALO) is introduced. The PHIPS-HALO combines imaging technique with simultaneous measurements of the angular light scattering, and can be operated in combination with the SID-3 instrument to investigate the effects of both small- and large-scale complexity to the angular scattering function. These instruments were operated in cloud simulations chambers as well as – in case of the aircraft probes – onboard three different research aircrafts. The cloud simulation chambers as well as the airborne platforms are described at the end of this chapter.

3.1. The scattering and depolarisation instrument SIMONE

The scattering and depolarisation instrument SIMONE[1] (Schnaiter et al., 2012) was designed at the Karlsruhe Institute of Technology to investigate the link be-

[1] SIMONE is an acronym for the German project name *Streulichtintensitätsmessungen zum optischen Nachweis von Eispartikeln*, which can be translated as Scattering Intensity Measurements for the Optical Detection of Ice Particles

tween microphysical properties of small ice particles and their near-backscattering linear depolarisation ratio, δ_L. The first version of the instrument is installed at the aerosol and cloud simulation chamber AIDA[2] of Karlsruhe Institute of Technology, where it has been operational at its current state since 2006. After installation, SIMONE has been successfully used in optical studies to investigate the depolarisation properties of laboratory produced ice particles (Schnaiter et al., 2012; Järvinen et al., 2014) and airborne dust particles (Järvinen et al., 2016a). Furthermore, the instrument has been employed in several ice nucleation campaigns to detect the onset of ice nucleation (e.g. Möhler et al., 2008; Steinke et al., 2011; Wagner et al., 2011; Hiranuma et al., 2014).

In the context of this PhD project, I developed and build a new version of the SIMONE instrument to be operated at the CERN CLOUD simulation chamber. This new version, from now on referred as SIMONE-Junior, has the same operational principle as the first SIMONE, but I made improvements to achieve a better sensitivity to small depolarisation ratios and I performed modifications to be able to measure circular depolarisation ratios. I operated the SIMONE-Junior instrument in two CLOUD-campaigns between 2013 and 2014. In these campaigns, the SIMONE-Junior was a key instrument to define the onset of cloud formation and the cloud duration in aqueous phase chemistry experiments (Hoyle et al., 2016). In experiments with ice clouds the instrument was used to study the linear and circular depolarisation properties of small ice particles, and it provided a reference for single particle polarisation measurements (Nichman et al., 2016). The improved sensitivity of the instrument made it possible to detect small depolarisation ratios of only few percentage. This ability was used, for the first time, to optically detect the viscous phase state of SOA particles without sampling the SOA particles or removing them from the surrounding gas (Järvinen et al., 2016b). Since end of 2014, the SIMONE-Junior is installed at the AIDA chamber and it was operated alongside the original SIMONE instrument during the RICE03 campaign to investigate the link between ice particle complexity and their depolarisation properties and during the PÖLY01 campaign to investigate the dust depolarisation properties. After finishing my PhD work, the SIMONE-Junior instrument will return to the CERN CLOUD chamber as a part of the basic chamber instrumentation.

In the following sections, we will first look at the measurement principle of the SIMONE instruments and go through the modifications made for SIMONE-Junior. I will then explain the calibration procedures for balancing the gains

[2] AIDA stands for Aerosol Interaction and Dynamics in the Atmosphere

of the backward polarised and cross-polarised channels and how to align the optical components. The instrument description is largely based on already published material in Järvinen et al. (2016a) and Järvinen et al. (2016b).

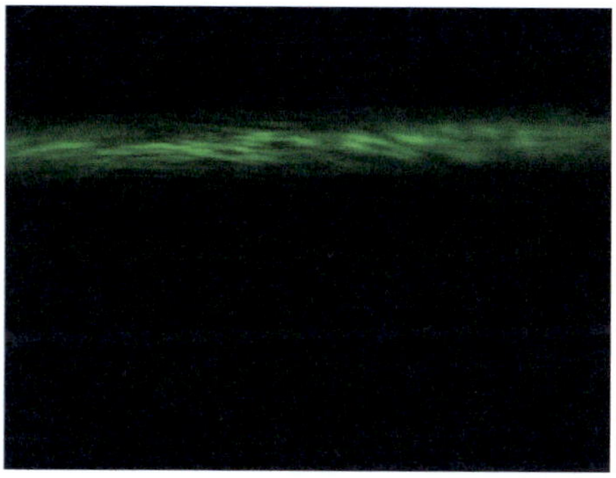

Figure 3.1.: Ice particles in the SIMONE-Junior 552 nm laser beam.

3.1.1. Measurement principle and setup

The SIMONE instruments detect light scattered by aerosol or cloud particles in near-forward (2° for AIDA chamber and 4° for the CERN CLOUD chamber) and near-backward (178° for the AIDA chamber and 176° for the CERN CLOUD chamber) direction. The first SIMONE version, hereafter called as SIMONE-Senior, uses a linearly polarised continuous wave (cw) semiconductor laser emitting at 488 nm wavelength. The polarisation vector of the incident light beam can be controlled with a liquid crystal polarisation rotator (LPR), and is usually aligned parallel or perpendicularly to the scattering plane. In the backward direction, the depolarisation of the incident linear polarisation state is determined by decomposing the backward scattered light into its vertically and horizontally polarised components. SIMONE-Junior has the same operation principle as SIMONE-Senior but a few changes were made to allow the production and measurement of circularly polarised light. These changes are briefly described in the following.

A schematic image of the main components of SIMONE-Junior is shown in Fig. 3.2. The incident light is produced with a 552 nm linearly polarised cw laser (OBIS 552 LS, Coherent, Inc.), and similarly to SIMONE-Senior, the laser polarisation vector can be rotated with a LPR (LPR-100, Meadowlark Optics). Additionally, a liquid crystal variable retarder (LVR-100, Meadowlark Optics) was introduced into the system. This variable retarder (LVR) can be set to have a retardance of zero, so that the light polarisation state stays unchanged, or it can be operated as a quarter wave plate with fast axis set to 45° with respect to the incident linear polarisation. The latter setting allows the production of circularly polarised light. The polarisation state of the laser can be changed between linear and circular with a response time of two seconds. Usually, the instrument is operated by alternating between right-handed circular polarisation and linear polarisation with a polarisation vector set parallel to the scattering plane. Before entering the cloud chamber, the laser beam is passed though a beam expander (BE10M-A - 10X, Thorlabs, Inc.) with tenfold magnification. The beam expander reduces the divergence of the laser to 0.002° (0.07° for the OBIS laser), which results in a more collimated laser beam at the centre of the chamber. The characterisation of the beam expander is described in Appendix A.

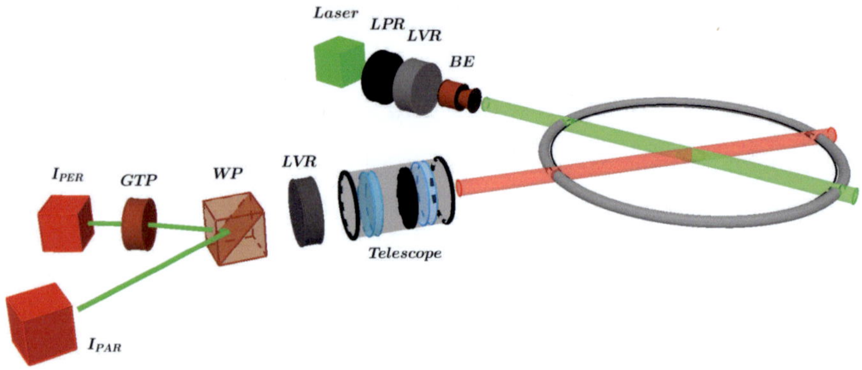

Figure 3.2.: A schematic illustration of the SIMONE-Junior operation principle.

The backward scattered light is passed through a telescope that collects light from the centre of the chamber. The telescope optics are composed of of an objective/ocular lens pair with focal lengths of 200 mm and 50 mm, respectively. The position of the lenses are chosen so that their focal points are coinciding at a pinhole. The size of the pinhole, 0.3 mm, was chosen so that the divergence angle of the telescope detection is defined to be 0.043° and the intersection between the laser beam and the detection apertures defines a detection volume of approximately 30 cm^{-3} (see Appendix A for details). The detection volume is large enough that it always occupies multiple particles under typical experimental conditions, and, therefore, the measured scattering signal is an average of a particle ensemble.

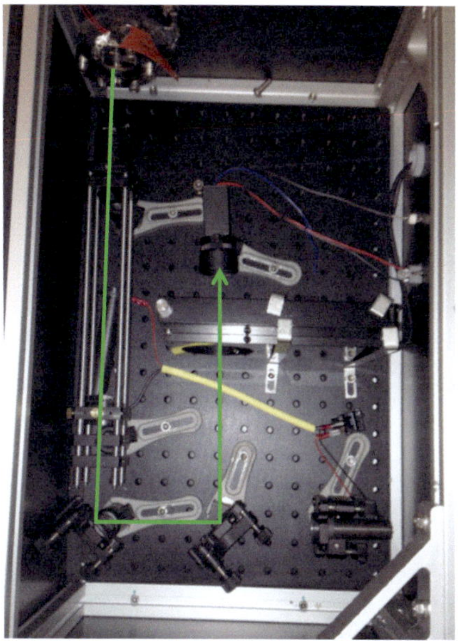

(a) SIMONE-Junior forward detection.

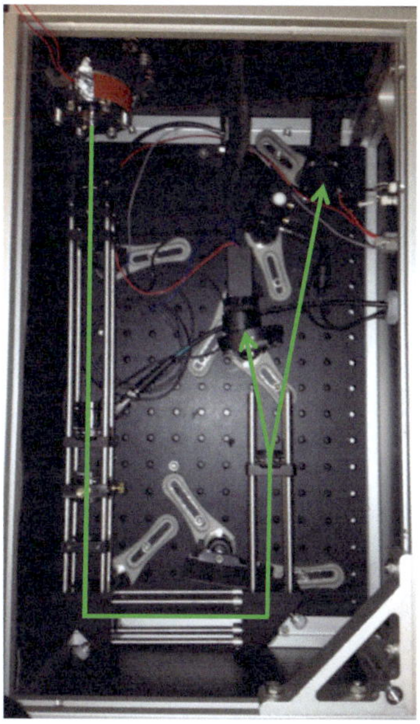

(b) SIMONE-Junior backward detection.

Figure 3.3.: Photograph of the SIMONE-Junior forward (a) and backward (b) detection after assembly of the optical components was finished. The green arrows illustrates the path of the collected rays.

Before analysis, the collected light is passed through a second liquid crystal variable retarder (LVR). In the case of circular incident polarisation, this LVR is operated as quarter wave plate with reversed orientation with respect to the first LVR, i.e. the second LVR changes the circular polarisation back to linear. In the case of linearly polarised incident light, the second LVR is set to zero retardance, and the polarisation state of the scattered light will remain unchanged. After this, the backscattered light is decomposed by a Wollaston prism (WP10-A, Thorlabs, Inc.) according to its parallel and perpendicular components with respect to the incident laser polarisation. The corresponding intensity components $I_{\|}$ and $I_{\perp}$ are measured by two photon multipliers (H10682-210, Hamamatsu) equipped with interference filters to reduce stray light noise. An additional filter flip with a 10% attenuation filter (optical thickness of 1) was installed in front of the detector for the polarised channel to avoid saturation of the signal. A Glan-Taylor prism was installed in front of the detector for the cross-polarised channel ($I_\perp$) to remove residual crosstalk intensity induced by the Wollaston prism.

The forward scattered light is collected with an identical telescope setup and guided using two mirrors to a photon multiplier (H10682-210, Hamamatsu) to measure the total scattered intensity (Fig. 3.3a). A filter wheel equipped with 10% and 100% attenuation filters (optical thickness 1 and 2, respectively) was installed in front of the photon multiplier to avoid saturation of the signal in case of a dense cloud. The exact optical thicknesses for the attenuation filters were characterised and are listed in Appendix A. The telescope at the forward scattering direction can be illuminated from behind with an additional guiding laser (CPS635R, Thorlabs, Inc., 635 nm). This guiding laser is used to align the field of views of the two telescopes so that they are facing each other. The alignment procedure using the guiding laser and a Spectralon target is described in Sec. 3.1.2.

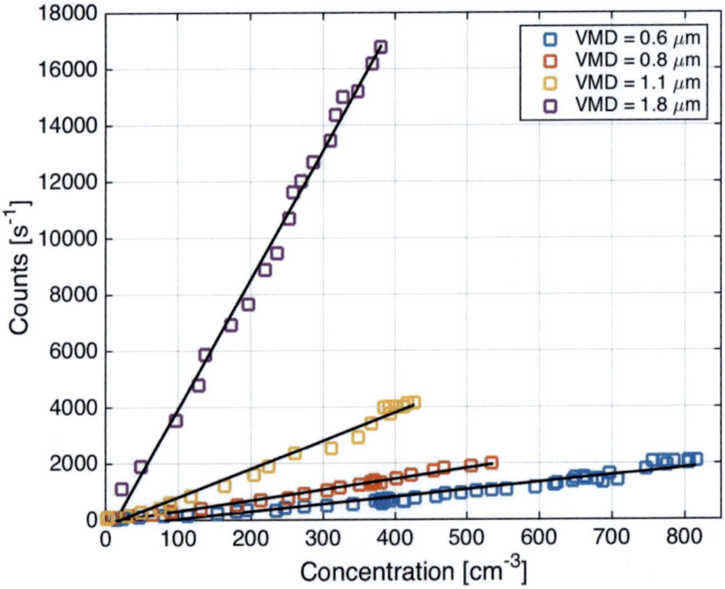

Figure 3.4.: The intensity of scattered light measured with the SIMONE-Junior backward detectors at an angle of 178° as a function of the particle concentration. The different symbols represent different particle ensembles with volume mean diameters (VMD) of 0.6 µm (blue squares), 0.8 µm (red squares), 1.1 µm (yellow squares) and 1.8 µm (purple squares).

It is important that the measurements are not affected by multiple scattering, so that single scattering approximation can be used. If the single scattering assumption is valid, the scattered light intensity is directly proportional to the amount of particles in the volume (Hovenier et al., 2003). I tested the single scattering assumption by dispersing dust particle ensembles with different geometric mean diameters into the chamber. Figure 3.4 shows the scattering intensity at the backward detectors of SIMONE-Junior as a function of the particle concentration during dispersion of four particle ensembles with different median diameters. We see that at the typical concentrations used in the chamber experiments, a linear dependency is always observed between the aerosol concentration and the scattering intensity. Therefore, our chamber measurements are not affected by multiple scattering and the single scattering approximation is valid.

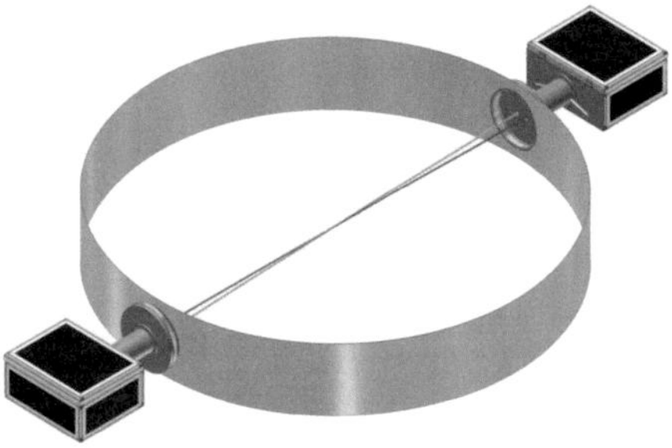

(a) A schematic of the mounted SIMONE-Junior instrument boxes. The green line represents the laser path and the purple line the telescope view.

(b) The inner flange that isolates the setup from the chamber volume and prevents contaminations from entering the chamber.

Figure 3.5.: SIMONE-Junior setup in the CERN CLOUD chamber.

Setup at the CERN CLOUD chamber

The optical components of the SIMONE-Junior instrument are mounted on two optical boards that are enclosed inside two light-tight aluminium casings (Figs. 3.3 and 3.5a). The forward-detection board hosts the forward detection unit and a beam dumb for destroying the laser beam (Fig. 3.3a) whereas the backward-detection board comprises the laser transmitter unit and the detection unit for backward scattered light (Fig. 3.3b). Both casings can be opened at any time for re-alignment of the optical components. The backside of the casings consist of stainless steel plates with two flanges for optically coated windows. These windows have a diameter of 48.3 mm and they are located 85 mm apart from each other. The windows isolate the optical components from the chamber volume, so that all components are kept at room temperature at all times. Due to this temperature gradient, all four windows are heated with foil heaters to avoid condensation or frost on their surface.

The two instrument cases of SIMONE-Junior were installed at the mid-level of the CERN-CLOUD chamber on two flanges facing each other (Fig. 3.5b). Between the chamber flanges and the casing there is a 1 m long stainless steel cylinder that can be evacuated to thermally isolate the instrument. The cylinders and the flanges are carrying the complete weight of the two casings (approximately 30 kg each) and no additional support is needed. This has the advantage that the instrument is adapting to any movements of the chamber walls.

Fig. 3.5a illustrates the SIMONE setup at the CERN CLOUD chamber. A laser beam (green line in Fig. 3.5a) is emitted from the laser side and it is guided to the laser destruction section at the other site. The optical axes of the two telescopes are illustrated with a purple line that crosses the laser beam at the centre of the chamber. The forward detection and the backward detection are facing each other and are located at an offset angle from the exact backward and forward directions that is defined by the diameter of the chamber. Using the geometry of the CERN CLOUD chamber, I derived a measurement angle of 176° for the backward detection and 4° for the forward detection with about 0.5° uncertainty (see Appendix A for derivation).

Setup at the AIDA chamber

The two SIMONE-Junior instrument casings were installed at the second level of the AIDA chamber replacing the SIMONE-Senior instrument on that position. Fig. 3.6 shows a schematic image of the installation setup on the second AIDA level. Unlike the situation at the CERN CLOUD chamber, at the AIDA chamber the instrument is mounted on flanges that are an extension of the chamber beyond the thermal insulation. This means that only one set of windows isolate the optical components from the chamber volume, and like in the CERN CLOUD setup, these windows are heated with foil heaters. The advantage of this setup is that condensation on the additional inner windows can be avoided and the laser beam is less affected by additional refraction and scattering processes on the surfaces of these windows. The sampling volume and the angles of detection are again defined by the geometry of the chamber. As the AIDA chamber has a larger diameter than the CERN CLOUD chamber, the depolarisation detection is closer to the exact backward direction, at an angle of 178°, and, consequently, the forward scattered light is detected at 2° angle (Schnaiter et al., 2012). A scattering target is permanently installed inside the AIDA chamber. This target is located at the centre of the chamber, just above the sampling volume of the instrument. For alignment and calibration purposes the target can be lowered so that the laser beam and the optical axes of the telescopes are penetrating the target at near-normal angles.

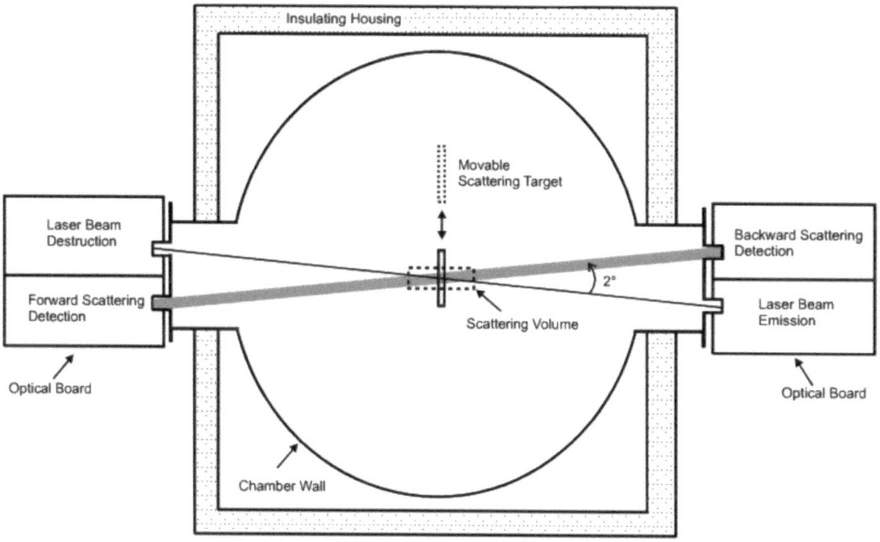

Figure 3.6.: Horizontal cross section of the second level of the AIDA chamber, where SIMONE-Junior was installed. Adapted from Schnaiter et al. (2012).

3.1.2. Alignment of the instrument and correction of the photon multiplier gains using a scattering target

As already mentioned, the two detection apertures of the SIMONE instrument are designed to face each other, which can be used to align the detection geometry of the instrument. For the alignment, the aperture of forward telescope can be visualised by illuminating its ocular lens with a guiding laser. Then, the positions of the two telescopes are adjusted until the light from the guiding laser passes completely through the objective lens and the pinhole of the backward telescope. After this, the laser is guided to overlap with the detection apertures at the centre of the chamber. This overlap can be visualised by moving a scattering target into the optical axes, so that both the laser beam and the optical axes of the telescopes are penetrating it (Fig. 3.7b). The laser beam is adjusted until there is a maximum overlap between the laser beam cross section and the apertures of the detection telescopes. After this, the positions of the mirrors, the polarisation beamsplitter and the photon multipliers are aligned by following diffusely scattered laser light from the target through the optical components of the forward and backward detection units.

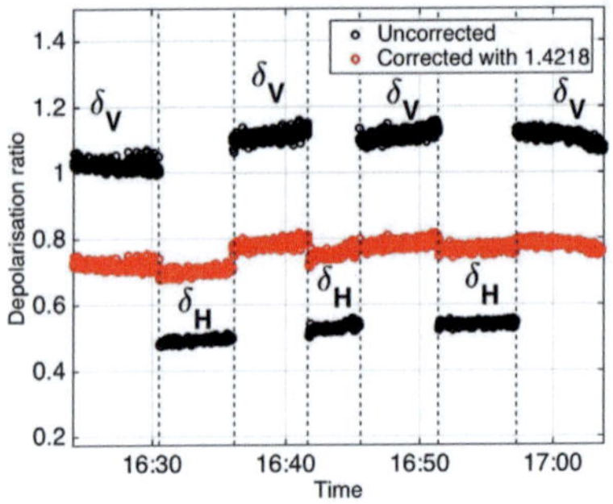

(a) Adjustment cycle for the backward photon multipliers using the scattering target.

(b) The overlap of the laser beam and the detection apertures on the scattering target.

Figure 3.7.: Alignment of the instrument.

The material of the scattering target is highly diffusing and depolarising Spectralon (SPO-SDM-200-SU, Laser 2000 GmbH) (Haner et al., 1999). The depolarising ability of the target can be used to align the gains of the two backward photon multipliers by defining a correction factor that is used to correct the signal of the cross-polarised photon multiplier in case of parallel incident polarisation. In case of perpendicular incident polarisation the same photon multiplier is detecting the polarised signal and, therefore, the correction is used in this case for the polarised channel. Besides the electrical gain, the correction factor also corrects the losses in the Glan-Taylor prism in front of the cross-polarised detector and differences in the interference filters installed in front of the photon multipliers. Fig. 3.7a shows an adjustment cycle performed before the PÖLY01 campaign, where depolarisation ratio of the illuminated Spectralon target was measured. The measurements with parallel and perpendicular incident polarisation are marked as δ_h and δ_v, respectively. After a correction factor of 1.4218 has been applied, we get the same depolarisation ratio of 0.76 for parallel and perpendicular incident polarisation. The correction factor might change within time and should be re-defined before every measurement campaign. Table 3.1 shows the correction factors measured for each campaign discussed in this thesis. Some difference in the correction factor is seen between measurements at the CERN CLOUD chamber compared to AIDA chamber, which is most likely due to different geometry of the chambers. The small variance in the correction factor within the measurements at the AIDA chamber can be used to estimate a relative uncertainty of 3% for the depolarisation measurements.

Chamber	Campaign	PM Gain Correction Factor
CERN CLOUD	CLOUD8	1.3479
CERN CLOUD	CLOUD9	1.3479
AIDA	RICE03	1.4908
AIDA	PÖLY01	1.4218

Table 3.1.: Table of photon multiplier gain correction factors defined before each measurement campaign.

3.1.3. Calibration procedure with spherical sulphuric acid solution droplets

Imperfect alignment and non-ideal behaviour of the optical components can induce crosstalk between the parallel and perpendicular channels in the backward scattering direction, i.e., in the perpendicular channel a percentage of the parallel intensity is measured and vice versa. This crosstalk ultimately determines the limit of detection (LOD) for the depolarisation ratio. The LOD can also be defined as the depolarisation crosstalk that is measured in the case of spherical particles.

To determine the LOD, I measured the depolarisation ratio of laboratory-generated sulphuric acid (SA) solution droplets. SA droplets are known to be spherical (e.g., Schnaiter et al., 2012), and since in the calibration I restricted to sub-micrometer sizes, the SA particles should not introduce linear or circular depolarisation at the measurement angle of SIMONE-Junior. Nevertheless, I measured a finite depolarisation signal, which represents the crosstalk from the parallel channel to the perpendicular channel. To minimise this depolarisation crosstalk, I optimised the alignment of the optical components by separately rotating the optical planes of each component until a minimum depolarisation signal was measured. Fig. 3.8 shows calibration curves attained by rotating the optical axis of the Wollaston prism. At the optimum alignment a minimum crosstalk depolarisation signal of 0.002 and 0.015 was measured for the linear (parallel) incident polarisation and for the circular (right-handed) incident polarisation, respectively. However, it has to be kept in mind that the alignment of the instrument may change over time, as the instrument is mounted on an expansion chamber. This also means that the LOD will increase from the calibration value. To be absolutely confident that a measured depolarisation signal is significant, I specified the LOD for linear depolarisation to be 0.01 and for circular depolarisation to be 0.02. If the depolarisation ratio is below these thresholds, I consider the particles to be non-depolarising according to my method.

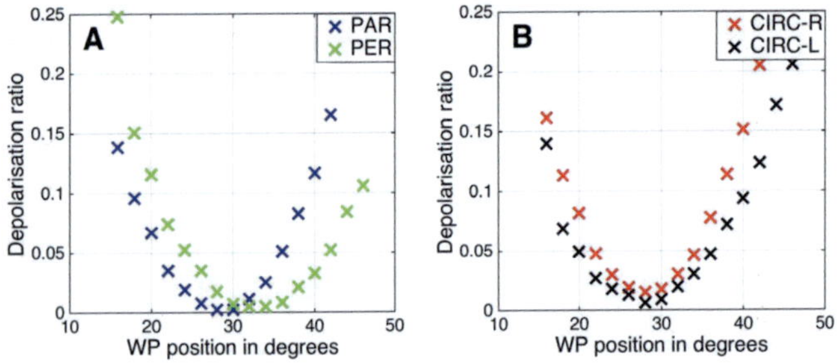

Figure 3.8.: The calibration curves for linear incident polarisation (A) and for circular incident polarisation (B).

3.1.4. Data analysis

The particle near-backscattering depolarisation ratio (δ_p) is defined as the ratio of the cross-polarised component to the polarised component (Sakai et al., 2010). To get this from our measurements, we need to subtract the contribution of the backscattering by air molecules and chamber walls. This background contribution is determined through a background measurement with a particle free chamber with the assumption that the addition of aerosol particles does not significantly change the characteristics of the molecular scattering or the chamber wall reflections. The background corrected signal is further corrected by taking into account the photon multiplier gain correction factor and by correcting filters, if applied. Now, the linear and circular particle depolarisation is defined as

$$\delta_p = \frac{cf \cdot (I_{\perp} - I_{\perp}^{bg})}{f_b \cdot (I_{\parallel} - I_{\parallel}^{bg})}, \tag{3.1}$$

with cf denoting the photon multiplier gain correction factor, f_b the filter correction for parallel channel, $I_{\parallel,\perp}$ denoting the measured intensities and $I_{\parallel,\perp}^{bg}$ denoting the background intensities. In the case of the linear depolarisation ratio, this value is expressed using the elements of the scattering matrix by

$$\delta_{L,h} = \frac{S_{11} - S_{22}}{S_{11} + 2S_{12} + S_{22}} \tag{3.2}$$

and

$$\delta_{L,v} = \frac{S_{11} - S_{22}}{S_{11} - 2S_{12} + S_{22}}, \tag{3.3}$$

depending on the incident polarisation (h and v for parallel and perpendicular incident polarisation, respectively). For circular incident polarisation, the circular depolarisation ratio can be expressed using the elements of the scattering matrix by

$$\delta_{C,r} = \frac{S_{11} + S_{44}}{S_{11} - S_{44}} \tag{3.4}$$

and

$$\delta_{C,l} = \frac{S_{11} - S_{44}}{S_{11} + S_{44}}, \tag{3.5}$$

where indices r and l denote right-handed and circular left-handed circular incident polarisation, respectively. The full derivation of Eqs. 3.1-3.5 can be found in Appendix B. Normally, the measurements are conducted either with a parallel incident polarisation or circular right-handed polarisation, so in the rest of this thesis $\delta_{L,h}$ will be simplified as δ_L or δ and $\delta_{C,r}$ as δ_C.

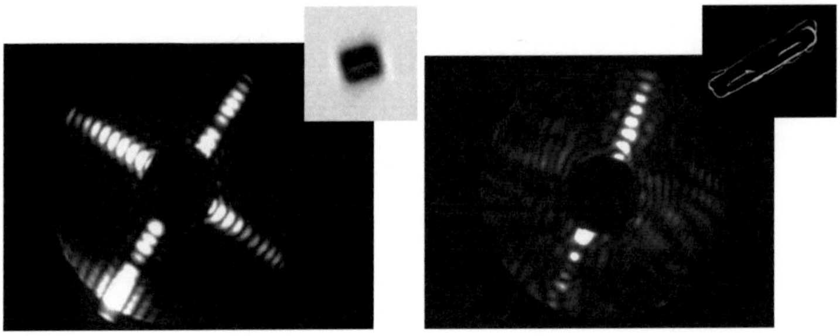

(a) A pristine column (b) A roughened column.

Figure 3.9.: Two examples of SID-3 scattering patterns of a pristine column (a) and a complex column (b).

3.2. Detection of particle small-scale complexity using 2D diffraction patterns

The information of the cloud particle geometry is traditionally relied on measurements with the Cloud Particle Imaging (CPI) probe (e.g. Connolly et al., 2005; Lawson et al., 2006, 2010; Schmitt and Heymsfield, 2014; Stith et al., 2014). The optical resolution of CPI probes is typically around 2 µm, which is enough to recognise the overlying geometry and to detect large-scale complexity, like aggregation. However, fine details of particle geometry is difficult to recognise using imaging techniques, especially in the case of small particles (Ulanowski et al., 2004). A solution to circumvent the optical resolution limitations of imaging probes is, instead of images, to acquire light-scattering "patterns". These light scattering patterns are more sensitive to small-scale geometry and complexity of particles, as is to be shown in the upcoming sections.

The University of Hertfordshire, UK, developed at the turn of the century a group of instruments to detect the forward scattering intensity of small cloud particles in angularly resolved manner. This new ability was applied to distinguish the forward scattering pattern of a droplet from an ice particle and, therefore, these instruments form the "Small Ice Detector" (SID) family. In the more modern versions of the SID instruments the angular resolution is increased to be able to record high-resolution scattering patterns. The newest version of the SID instruments is collectively called SID-3, which can record two-dimensional scattering patterns of single particles (Kaye et al., 2008). The SID-3 scattering patterns can be used to get detailed information on the particle microphysics: it is possible to recover the shape, size and orientation of the particles (Vochezer et al., 2016). Also, ice crystal small-scale complexity can be distinguished and quantified (Ulanowski et al., 2006, 2014; Schnaiter et al., 2016).

3.2.1. The measurement principle of the SID-3 and PPD-2K instruments

Here, the small-scale complexity is defined as all sorts of deviations from the pristine crystal shape that cause similar uniform spatial distribution of the scattered light. These deviations comprise ice particle surface roughness, hollowness, air bubbles within the crystal or polycrystals. The first investigations of the effects of small-scale complexity to angular light scattering was made with laboratory produced ice analogue particles (Ulanowski et al., 2006). It was found that a scattering patterns of a roughened hexagonal column showed "speckles", whereas the scattering patterns of a smooth hexagonal column had sharp edges and well defined intensity regions (Fig. 3.9). The number and distribution of the speckles could be related to the degree of the crystal complexity. Later, this method was calibrated in a set of cloud chamber experiments for cirrus ice particles (Schnaiter et al., 2016).

In this section, we will get familiar with the measurement principle of the SID-3 instrument and its laboratory version, the Particle Phase Discriminator Mark 2, Karlsruhe edition (PPD-2K). I present a method to discriminate between spherical and non-spherical particles, which can be used to detect ice particles in mixed-phase clouds (Vochezer et al., 2016) or optically spherical ice particles in convective outflows (Järvinen et al., 2016c). Later, the analysis of small-scale complexity is presented.

The SID-3 and PPD-2K record two-dimensional scattering patterns of single particles using a 780×582 pixels and a 582×592 pixels ICCD (Photek Ltd, UK) camera, respectively. As light source both of the instruments use a frequency-doubled Neodym YAG (Nd:YAG) laser that emits linearly polarised light with 100 mW power at a wavelength of 532 nm. With help of a quarter wave ($\lambda/4$) plate the linearly polarised light is covered into circularly polarised in order to minimise polarisation-dependent variations in the scattering patterns. A schematic illustration of the measurement principle of the SID-3 and PPD-2K instruments can be found in Fig. 3.10. If a particle passes the sensitive area of the laser beam, light scattered in the forward angular range from approximately 7° to 23° (in PPD-2K from 5° to 26°) is collected and an image of this annulus region is generated on the CCD chip of the camera. A beam dump destroys the laser beam at the near-forward direction to prevent the camera from being continuously saturated in this angular region and to reduce the stray light inside the detection optics.

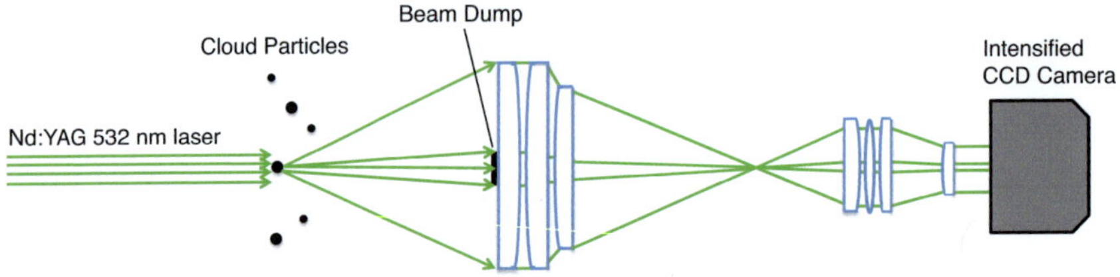

Figure 3.10.: Schematic image of the SID-3/PPD-2K measurement principle.

In the SID-3 the particle detection takes place along the laser beam that is guided vertically across the instrument head (Fig. 3.11a). The instrument head is designed with "open path"-principle, which means that the cloud particles are detected as they pass the laser beam without sampling them through a particle inlet. This reduces particle shattering that might happen at the edges of inlets. The sensitive area of the instrument is defined by the nested field of views (FOV) of two trigger detectors. The trigger detectors are aligned symmetrically along the laser beam axis and have a circular aperture with a half angle of 9.25° at 50° relative to the forward direction. The two trigger detectors have a different sized FOV on the flat $1500\,\mu m \times 160\,\mu m$ cross section of the laser beam, so that the FOV of trigger detector one lies within the FOV of trigger detector two. A particle is detected only if it is in the FOV of both trigger detectors, i.e. inside the sensitive area of the instrument. In order to determine the sensitive area of the SID-3 trigger detectors I used a piezo electric droplet generator (GESIM GmbH, Grosserkmannsdorf, Germany) producing droplets of $80\,\mu m$ in diameter mounted on a x-y-z -stage to map the FOV of the two trigger detectors. Fig. 3.11b shows the results of this mapping. The FOVs of the trigger detectors are projected as ellipsoids, so that the smaller FOV (red crosses) is located inside the larger FOV (black crosses). I defined the sensitive area as the area of the smaller FOV taking into account the droplet dimension (green crosses). An integral over this area gives a sensitive area of $0.47\,mm^2$. In the PPD-2K, the sample flow is focused on the laser beam, so that all the particles in the sampling flow can be detected. The focusing of the sample flow in PPD-2K leads to better counting statistics compared to the SID-3.

The sensitive area of the SID-3 is located about 2 cm from the upper part of the instrument (Fig. 3.11a). At this position the laser beam has a power of 30 mW (measured with a energy sensor, EnergyMax-RS J-50MB-YAG, Coherent). The trigger system detects particles in the sensitive area, after which the scattering pattern is recorded by the CCD camera. The CCD camera in both instruments has a maximum repetition rate of 30 Hz. This means that particles arriving during the dead time of the camera will not be imaged. In both of the instruments, the trigger electronics have a higher repetition rate of 11 kHz, so that the trigger information is stored, even if the camera is busy. In the SID-3 the trigger information is stored as a trigger intensity histogram with a predetermined bin width, whereas in the PPD-2K every triggered particle is stored separately. In both cases, the trigger information can be used to calculate the particle size distribution.

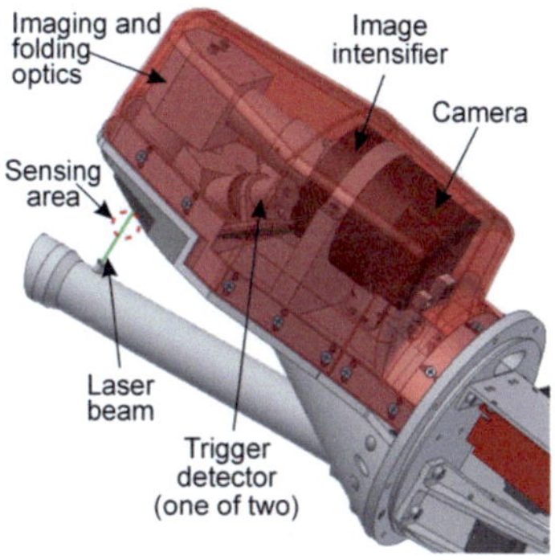

(a) SID-3 instrument head. Figure from Ulanowski et al. (2014).

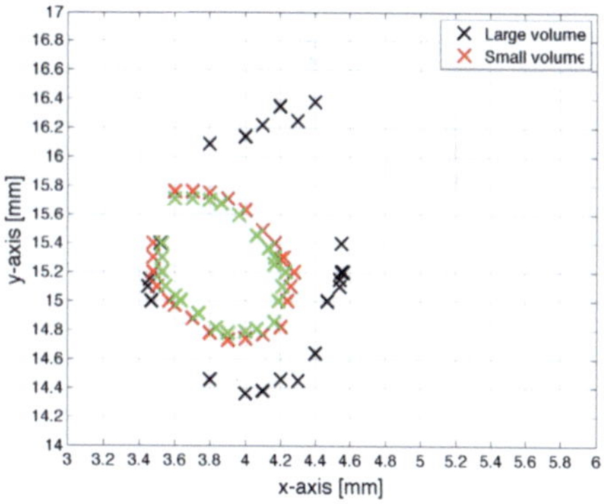

(b) A measurement of the SID-3 sensitive area

Figure 3.11.: The SID-3 setup (a) and the sensitive area of the instrument (b).

Calculating particle size distribution from trigger information

The trigger intensity can be related to the particle size through the following equation (Cotton et al., 2010)

$$D_p = a \times I^b, \tag{3.6}$$

where D_p is the particle diameter and I the trigger intensity. The pre-factor a is a function of the laser power and of the gain applied to the trigger detector which can be adjusted. The exponent b is a constant obtained from a fit to simulated scattered irradiances for a circular aperture with a solid angle of the SID-3 or PPD-2K trigger detector calculated by Mie theory for water droplets (Vochezer et al., 2016). For the SID-3, the trigger detector has a circular aperture with a half angle of 9.25° at 50° relative to the forward direction, which leads to an exponent $b = 0.517$. In the case of the PPD-2K the trigger detector covers 7.4° to 25.6° relative to the forward direction and the exponent b is consequently 0.522.

The pre-factor a can be derived from a calibration measurement with water droplets. The exact size of the droplets can be determined by using a two-step Mie fit routine to the imaged scattering pattern (Vochezer et al., 2016). In the subsequent calibration procedure the exact droplet size is compared to its trigger intensity. In the PPD-2K the trigger intensity corresponds to the peak height of the trigger peak, so the trigger intensity is only dependent on the applied gain. In the experiments presented in this thesis I used gain of 60, which leads to a pre-factor $a = 0.885$. In the SID-3 the trigger intensity is an integrate over the trigger peak, and therefore, depends also on the particle speed in the laser beam: particles with lower speeds induce a wider peak and, therefore, a larger integrated intensity compared to faster particles. Therefore, a laboratory calibration with particles traveling at $10\,\mathrm{m\,s^{-1}}$ is not valid for aircraft measurements, where the particle speeds can range from $80\,\mathrm{m\,s^{-1}}$ to $250\,\mathrm{m\,s^{-1}}$ depending on the aircraft. Fig. 3.12 shows the calibration curves for laboratory experiments (solid curve) and for aircraft studies (dashed curves). We see that the pre-factor a increases with increasing particle speed. Therefore, the pre-factor a was determined separately for each measurement campaign using droplet data recorded during those campaigns.

To derive particle number concentration the particle counts are divided by the sampling volume. In the SID-3 the sampling volume is a column defined by the sensitive area and the instrument speed with respect to the particles. At an airspeed of $200\,\mathrm{m\,s^{-1}}$ a volume of $94\,\mathrm{cm^{-3}}$ is sampled every second, whereas in the laboratory with $10\,\mathrm{m\,s^{-1}}$ airspeed the sampled volume per second is only $4.7\,\mathrm{cm^{-3}}$. In PPD-2K the sampling volume is comparable to the volume of the airflow that passed through the instrument. In both instruments, the active sampling volume is reduced by the trigger electronic dead time of $8\,\mathrm{\mu s}$ (SID-3) and $8.25\,\mathrm{\mu s}$ (PPD-2K) (Johnson et al., 2014). Assuming cloud particle number concentrations of 20 to 300 cm^{-3} and a typical flow speed for the SID-3 of 100 m s^{-1} and a PPD-2K sampling flow of 5 l min^{-1}, we get a reduction in sample volume of 0.8 to 11.3 % (for the SID-3) and 1.4 to 20.6 % (for the PPD-2K) (Vochezer et al., 2016). The electronics dead time is corrected in the derivation of the number concentration.

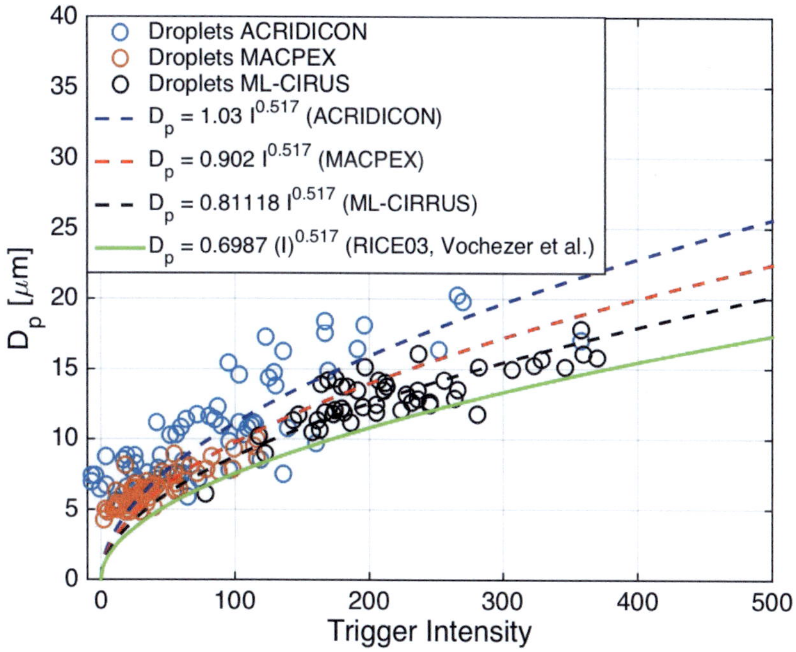

Figure 3.12.: SID-3 droplet calibration for different platforms

3.2.2. Determining particle sphericity

We can discriminate between spherical and non-spherical particles based on the azimuthal symmetry of their scattering patterns. The scattering pattern of a sphere shows concentric rings with a perfect azimuthal symmetry, whereas the scattering pattern of an aspherical particle has a lower azimuthal symmetry (Fig. 3.13a). The individual scattering patterns can be automatically analysed using a LabView (National Instruments, Inc., USA) based software developed at the Karlsruhe Institute of Technology (Vochezer et al., 2016). For each scattering pattern, the software first chooses a region of interest (ROI; indicated with green lines in Fig. 3.13a). Then, a saturation ratio $q = N_s/N_a$ and a mean intensity $\overline{I} = (\sum_{i,j}^{N,M} I(i,j))/N_a$ for the ROI are calculated. Here N_s is the number of saturated pixels, N_a is the number of all pixels, and $I(i,j)$ is the grey level intensity of pixel i, j in the ROI. The analysis is continued further if saturation ratio q is smaller than 0.15 and the mean intensity $\overline{I}$ larger than 3. These values are chosen to avoid images that are too bright or have a significant contribution of background noise.

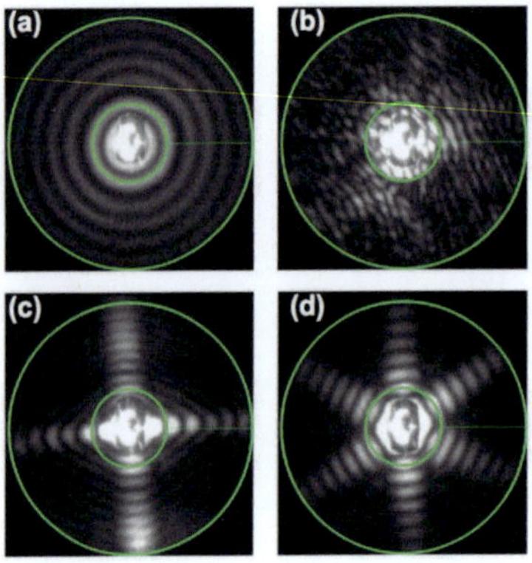

(a) PPD-2K scattering pattern of a) a droplet, b) a complex ice particle, c) a column with 4-fold symmetry and d) a plate with 6-fold symmetry.

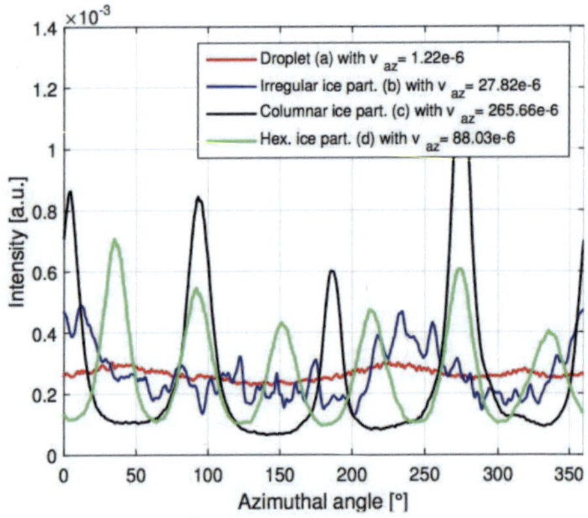

(b) Polar integrated azimuthal intensity profiles of the particles in Fig. 3.13a.

Figure 3.13.: The azimuthal profile of particles measured with the PPD-2K. Adapted from Vochezer et al. (2016).

In the next step, an unwrapped version of the ROI is generated, which is equivalent to a change from polar to Cartesian coordinates and carried out by a bilinear interpolation. The resulting unwrapped image is integrated over the polar angle to produce azimuthal profiles of the patterns as shown in Fig. 3.13b. The degree of symmetry in these profiles is determined by calculating the variance along the intensity profile

$$v_{az} = \frac{\sum_{i=1}^{N}(x_i - \mu)^2}{N-1}, \tag{3.7}$$

where N is the number of discrete angles of the azimuthal profile, x_i the value of a certain element, and μ the mean intensity value. A threshold value for spherical particles of v_{az}^{thr} of 1×10^{-5} was determined based on laboratory experiments with droplets and ice particles (Vochezer et al., 2016). This threshold can be used to determine aspherical fractions so that particles having a $v_{az} > v_{az}^{thr}$ are classified as aspherical.

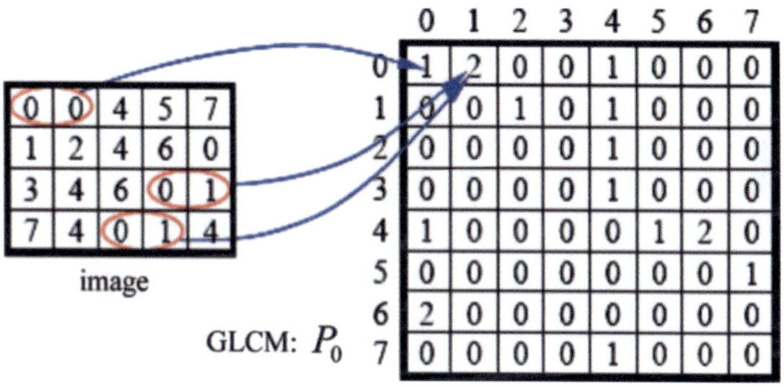

Figure 3.14.: An example of calculating a GLCM. From Lu et al. (2006).

3.2.3. Quantifying the degree of particle small-scale complexity

The analysis of the particle small-scale complexity from the scattering patterns relies on the grey-level co-occurrence matrix (GLCM) method described in Lu et al. (2006). This method was originally developed for in-process quality control of surface textures, but Ulanowski et al. (2010) transferred the method for examining surface roughness from SID-3 images. Later, the method was applied for studying ice crystal surface roughness in atmospheric ice clouds (Ulanowski et al., 2014) and, recently, Schnaiter et al. (2016) calibrated the method for cirrus ice particles in a series of laboratory experiments.

The GLCM describes how often pairs of grey-levels occur in the texture image for pixels separated by a certain distance and along a certain direction (Haralick et al., 1973). Fig. 3.14 shows an example of how the GLCM is calculated. The matrix in Fig. 3.14 has 8 grey values so that the resulting GLCM has 8×8 elements. The neighbouring pixel values are investigated and the number of pair combinations are filled in the corresponding elements of the GLCM. In the case of 8 bit grey-level scattering patterns imaged by the SID-3 instrument, the GLCM has a size of 128×128 elements.

In Fig. 3.14, the comparison was conducted along the horizontal direction. The same procedure can be done along other directions (or angles) as well as for other pixel distances, which would produce a new GLCM for each direction or distance. Usually, several GLCMs are calculated, since a single GLCM might not be enough to describe the textural features of the input image. From the GLCM several texture features can be calculated to define the surface structure (Haralick et al., 1973). Normally, five features (contrast, correlation, energy, entropy, and homogeneity) are computed, but Lu et al. (2006) added a new feature, the normalised energy feature parameter k_e.

The k_e parameter is defined from the energy feature that represents the sum of the squared elements in the GLCM

$$E = \sum_{i=0}^{m-1} \sum_{j=0}^{m-1} p(i,j)^2 \tag{3.8}$$

where $p(i,j)$ represents a specific pair of grey-level values given by the two matrix indices i, j. Now the k_e parameter can be obtained by fitting the energy feature curves of three different directions with the following curve equation

$$y = y_0 + k_e exp(-x/\sqrt{\sigma}) \tag{3.9}$$

where x indicates the offset distance d and y is the energy feature value.

Lu et al. (2006) defined that k_e was the most robust to variations in setup configuration. The same conclusion was made by Schnaiter et al. (2016), but now for analysis of the SID-3 scattering patterns. Therefore, for the rest of this work, I parameterise the particle small-scale complexity using k_e as the complexity parameter. The k_e was calculated for ice particles with mean intensity $\overline{I}$ between 10 and 25. This reduces any biases caused by intensity fluctuations and guarantees that only well illuminated particles are analysed.

As mentioned above, the complexity parameter k_e can be linked with an actual degree of particle complexity. Fig. 6.30 shows simulated SID-3 scattering patterns using a Gaussian random sphere model with different degrees of distortion represented with the distortion parameter σ (see details in Sec. 2.4.1). The k_e varies between 4 and 4.8 (in experiments the range is typically between 4 and 6) and increases with increasing complexity (σ). The dependence between the complexity parameter k_e and the simulated complexity is almost linear. This simulation illustrates that the complexity parameter k_e can distinguish differences in the particle complexity and, therefore, it represents a good measure for the small-scale complexity.

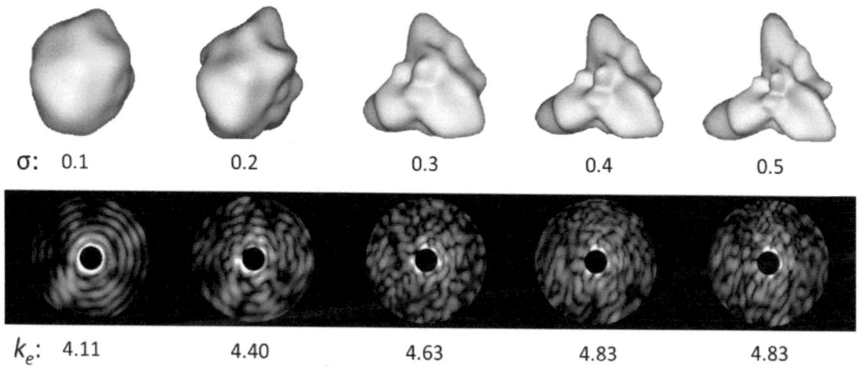

Figure 3.15.: Simulated SID-3 scattering patterns using a Gaussian random sphere model with different degrees of distortion. From Schnaiter et al. (2016).

Coincident particle imaging

In the case of SID-3, the camera FOV is 9 mm^2, which is large enough that multiple particles can occupy it simultaneously. If two or more particles are found in the FOV, the scattering patterns are affected. Examples of coincidence images are shown in Fig. 3.16. The scattering patterns from (a) to (c) are from coinciding droplets. Typically the second, out of focus particle leaves a shadow on the scattering pattern that can be seen as a "kidney shaped" feature (a) or as a circle of brighter and darker areas (b and c). Similar features can be observed in the case of ice particle patterns (d-f).
The probability of coincidence imaging was calculated in Vochezer et al. (2016). A coincidence particle imaging probability from 0.04 to 7.03% for particle number concentrations of 20 to 300 cm^{-3} was found and a coincident sampling probability of 1% was reached for a particle number concentration of 103 cm^{-3}. The coincidence imaging probability of a few percentages is significant and can, therefore, bias the SID-3 analysis. However, due to the high resolution of SID-3, the coincidence images can be manually identified (as done in Fig. 3.16) and removed from the analysis. I did manual cross-checks for all experiments, where typical particle concentrations were above 100 cm^{-3}. This mainly covers measurement flights in mixed-phase clouds (see Sec. 3.4.4). In the manual

cross-check, each scattering pattern was checked so that (i) droplets were not misclassified as ice particles and (ii) ice particles did not contain coincidence patterns. Only valid ice scattering patterns were accepted for the complexity analysis.

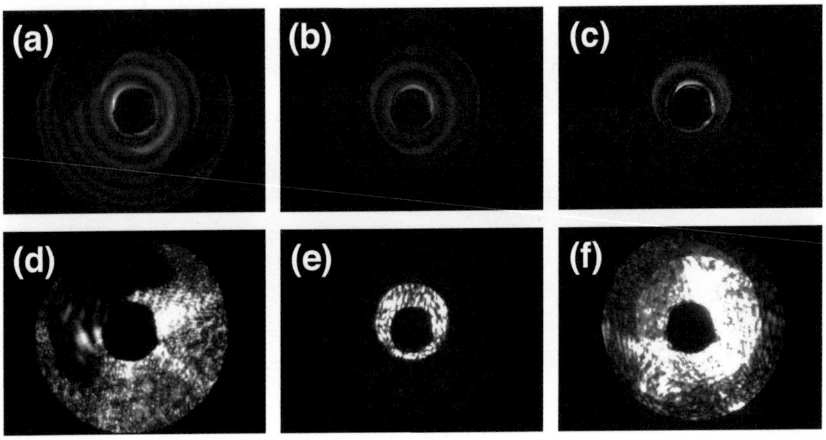

Figure 3.16.: Examples of coincidence scattering patterns. Particles (a)-(c) are droplets and (d)-(f) ice particles.

3.3. Simultaneous measurement of particle shape and their angular scattering function

There has been a lot of interest to combine a particle imager with a polar nephelometer for the simultaneous measurement of the angular scattering function of the imaged ice particle. Early versions of the CPI instruments tried to do this, but did not succeed to combine these two measurements in the same instrument (Paul Lawson, personal communication, 2015). Later, during the South Pole Ice Crystal Experiment a CPI instrument was combined for the first time with a polar nephelometer instrument hoping that the same ice crystals that has passed the CPI would be measured in the polar nephelometer installed behind the CPI outlet (Shcherbakov et al., 2006). The authors report a large variability in the angular dependency of the scattering intensities with single particles, also showing orientation dependent features, like specular reflections. To better quantify the effect of particle orientation and morphology on its single scattering properties was the motivation to develop a new instrument, where the particle imaging and scattering measurements are performed simultaneously on the same particle. This requires that the particle is imaged from two different views in order to determine its orientation in the laser beam. This idea was the beginning of the development of the Particle Habit Imaging and Polar Scattering (PHIPS) instrument. PHIPS was primarily developed for the HALO aircraft (see details in Sec. 3.4.3) with support from the Helmholtz Association (HGF), the Federal Ministry of Education and Research (BMBF) as well as the German Research Foundation (DFG). The project leader for the development was Dr. M. Schnaiter and the development was conducted in his group. The first prototype of the imager part of the instrument was build for the AIDA chamber (see details in Sec. 3.4.1) in the PhD project of R. Schön (Schön, 2007; Schön et al., 2011). The setup consisted of a single imaging system with a CCD camera and a broadband light source. Later, the setup was extended by a second imaging unit and by a polar nephelometer (PHIPS-AIDA; Abdelmonem et al. (2011)). Within his PhD project, Roland Schön developed an image analysis software (written in the Interactive Data Language, IDL) that is used to this day to determine the geometrical particle parameters from the PHIPS images, like projected area, maximum dimensions, aspect ratio, etc.

The aircraft version of the instrument, called PHIPS-HALO, is a successive development of these prototype instruments (Abdelmonem et al., 2016). Additional modifications in design and components were undertaken to prepare the instrument for operation on an aircraft platform. The PHIPS-HALO instrument was finished in 2013 and in the same year it was certified for the HALO aircraft. Within my PhD project, I characterised the polar nephelometer part of PHIPS-HALO and operated the instrument in the field campaigns ML-CIRRUS and ACRIDICON-CHUVA that were conducted in 2014. ML-CIRRUS was the depute for the PHIPS-HALO instrument and the first experience on a high-speed airborne platform. Although the acquired dataset from this campaign remained limited, we were able to improve the behaviour of the trigger electronics to ensure optimum performance also for airspeeds faster than $200\,\mathrm{m\,s^{-1}}$. The valuable experience from ML-CIRRUS was later a great benefit for the ACRIDICON-CHUVA campaign, and a large dataset was collected that is now a part of this thesis. Later in 2014, I operated the PHIPS-HALO instrument under AIDA in the Rough ICE (RICE03) campaign, where it run side by side with the established polar nephelometer (PN) from the University Blaise Pascal in Clermont-Ferrand, France.

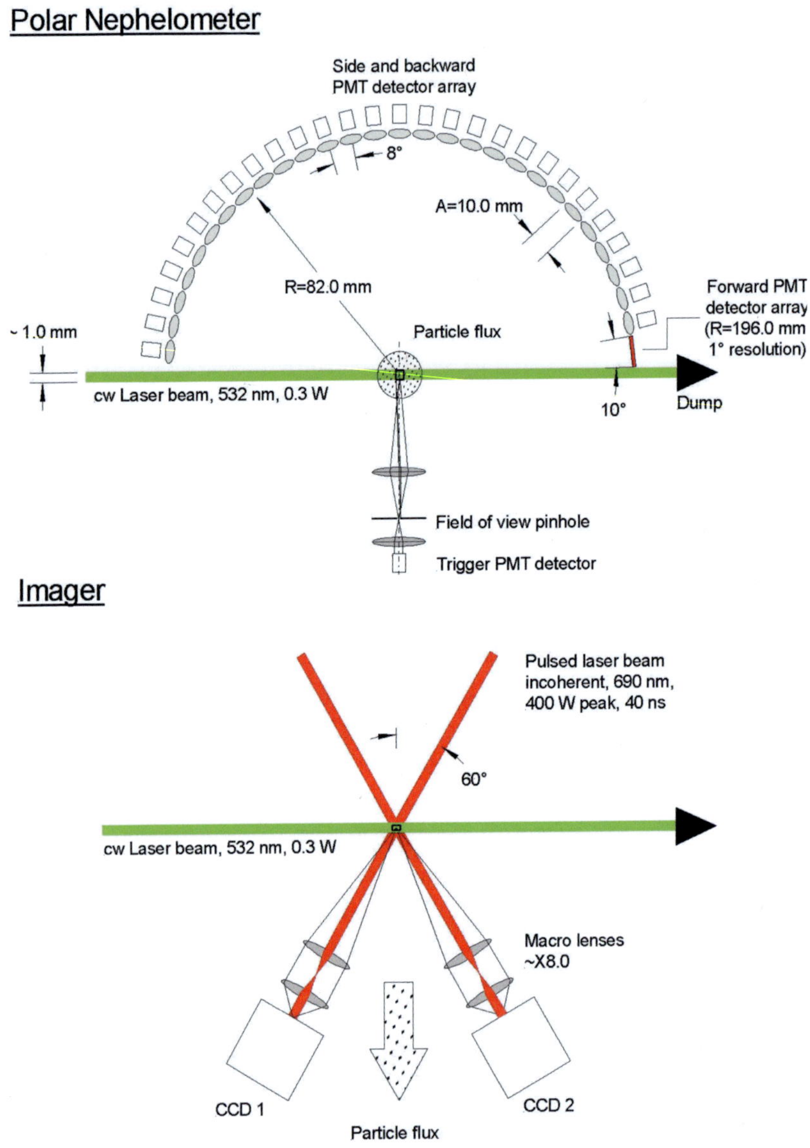

Figure 3.17.: PHIPS-HALO measurement principle. Adapted from Stegmann et al. (2016).

3.3.1. The measurement principle of the PHIPS-HALO probe

The basic measurement concept of PHIPS-HALO is simultaneous imaging of single ice crystal and measurement of their angular scattering function. The instrument can be divided into an imaging part and a polar nephelometer part that are combined by a trigger system (Fig. 6.5). The imaging part consists of two identical camera-telescope assemblies and a pulsed illumination laser. The two camera telescopes are placed so that they gather light scattered from the particle at angles of +60° and -60° from the central axis of the probe. The magnification of the camera telescopes can be varied from 1.4 to 9.0, which corresponds to a field of view range from 7.86 mm to 1.22 mm (diagonal size), respectively. In the ACRIDICON-CHUVA campaign, I used two different magnification settings for the two cameras to ensure that a particle event is always captured at least within one camera frame and – if well captured – a more detailed image can be found in the second camera frame.

The magnified bright field images that a gathered by the two telescopes are projected on the CCD chips of the cameras (GE1380, Allied Vision Technologies Inc., Canada), each having an image resolution of 1360x1024 pixels and a maximum frame rate of 20 Hz. This specific camera model was chosen because it has a well-defined shutter lag of 4 µs that enables exact timing for the illumination laser. The particles are illuminated using an incoherent pulsed diode laser (Cavitar, model Cavilux, 690nm, 400 W, Class 4). The use of incoherent laser light enables the production of diffraction- and chromatic aberration-free bright field microscopic images.

The polar nephelometer part of PHIPS-HALO measures the light scattered from particles as they pass through the horizontally aligned scattering laser beam. The scattering laser is a frequency doubled Nd:YAG laser (CrystaLaser, model CL532-300-L) with a wavelength of 532 nm and an output power of 300 mW. The detection volume is located 161 mm from the laser exit 20 mm above the optical plate and coincide with the central axis of the inlet tube. At this position, the laser is linearly polarised with a polarisation direction vertical to the scattering plane of the nephelometer. The divergence of the laser is 2 mrad (measured 1.5 mrad), which leads to a cross section of 1.02 mm (measured with a laser beam profiler, PowerMax USB - PS19 Power Sensor) at the position of the detection volume.

The light scattered from a particle is collected with 30 PMMA[3] fibres and coupled to a multi-anode photomultiplier array (MAPMT, Hamamatu, model H7260). This covers the angular range from 1° to 170°. The first 10 forward scattering PMMA fibres (1-10°) are placed at a scattering distance of 196 mm from the scattering centre, so that one degree angular resolution is achieved. The side- and backward scattered light is collected with 20 off-axis parabolic mirrors and is focused into the PMMA fibres. The parabolic mirrors are placed at equidistant angular separations of 8° (from 18° to 170°) at a scattering distance of 82 mm. Their diameter is 10 mm so that the angular range that each mirror cover is ±3.5°.

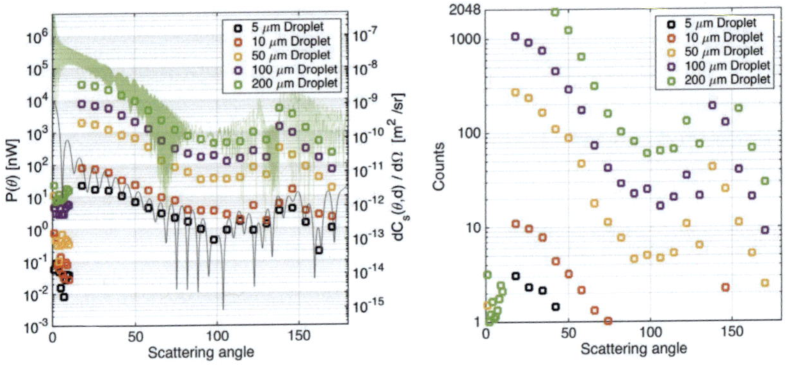

Figure 3.18.: (Left) Theoretical light power that reaches the detectors by laser light scattering on water droplets with diameters of 5 µm (black squares), 10 µm (orange squares), 50 µm (yellow squares), 100 µm (purple squares) and 200 µm (green squares). For 5 µm and 200 µm droplets also the scattering cross sections are shown in high angular resolution. (Right) Digitised particle scattering functions that are expected for a single MAPMT gain setting.

Modelling the scattering response of the polar nephelometer part

To determine the dynamic range of the polar nephelometer part of the PHIPS-HALO instrument, I modelled the response of the detectors to scattering of spherical particles. The total power that arrives in the solid angle of a detector from scattering of a sphere with known refractive index is dependent on the

[3] Poly(methyl methacrylate), also known as acrylic, acrylic glass or Plexiglas™

incident laser intensity, the distance of the detector to the scattering centre, and the detector geometry through the following equation

$$P(\Theta) = \frac{I_0}{4\pi r^2} \iint_A \frac{dC_s}{d\Omega}(d,\theta)d\Omega \tag{3.10}$$

where I_0 is the laser intensity, r the distance from scattering centre to the detector, Ω the solid angle of the detector and $\frac{dC_s}{d\Omega}(d,\theta)$ the differential scattering cross section of a particle with a diameter d. Since the detector areas are spherical for both the forward direction fibres as well as the parabolic mirrors in the sideward direction, the double integral 3.10 can be analytically solved with respect to the azimuthal angle, and the resulting integral depends only on the polar angle as follows

$$I = 2I_0 \int_{\theta_0-\alpha}^{\theta_0+\alpha} \frac{dC(\theta)}{d\Omega} arctan(\sqrt{tan^2(\alpha) - tan^2(\theta-\theta_0)})d\theta, \tag{3.11}$$

where α is the half opening angle of the detector with a radius of $R_{detector}$ $(arctan(R_{detector}/r))$. The scattering cross section of spherical particles can be calculated using Mie theory (for this the Bohren and Huffman code was used, Bohren and Huffman (2008)) and for vertically polarised incident light the differential scattering cross section can be expressed as following

$$\frac{dC_s}{d\Omega(d,\theta)} = \frac{|S_1|^2}{k^2}. \tag{3.12}$$

The theoretical power measured by a detector (nW) was calculated for five droplets with different diameters (refractive index of 1.34 and diameters ranging from 5-200 µm) and using the above mentioned specifications for laser power and the scattering geometry. The results of these calculations are shown in Fig. 3.18. The total scattered intensity measured at a single detector varies three orders of magnitude between the 5 µm and the 200 µm droplet. Also, the intensity measured at the 18° detector is two orders of magnitude higher than the intensity measured at 90° angle for the same droplet size. The calculation for the differential scattering cross section (black and green solid curves) show oscillation features that are typical for spherical particles. However, these features have a frequency that is significantly higher than the angular resolution of the instrument, and thus cannot be resolved from the measurements. Larger features, like the rainbow around 140° or the decrease in phase function around

100° for the 5 µm droplet and around 130° for the 200 µm droplet, can be detected.

Although the MAPMT has a high dynamic range, the analog-to-digital (AD) electronic processing of the scattering data restricts the resolution of the detection to 11 bits. I simulated the possible response of the electronics assuming ideal optical components with perfect transmission and no crosstalk. The particle size range of interest was chosen to be from 50 µm upward, since at larger particle sizes the imaging part of PHIPS-HALO will provide more detailed information of the particles. I set the intensity of the 90° detector for the 50 µm droplet to be equivalent of 5 counts (assuming that the MAPMT has a background of a few counts) and scaled the scattering responses of the other droplets respectively. This represents the fact that the MAPMT works only with one gain setting for all channels. From Fig. 3.18 we see that with the proposed gain setting we could detect the angular scattering function of the 50 µm and 100 µm droplets in the angular range of 18-170° while the scattering function of the 200 µm droplet can be measured from 42° onwards. In conclusion, we should be able to image and simultaneously measure the angular scattering function of particles from 50 µm up to almost 200 µm in diameter. It has to be kept in mind that with ice particles we can expect somewhat flatter angular scattering functions, which would increase the measurement range, when measuring in cirrus clouds. By adjusting the MAPMT gain, the measurement range can be shifted to smaller or larger particle size ranges.

The forward scattering detectors (1-10°) were positioned to a larger distance from the scattering centre to get one-degree resolution. The theoretical calculations, however, show that this decreases the intensity in the forward detectors to a degree that the dynamic range of the instrument is not high enough to measure these low intensities. To improve this, the forward channels would need a separate photon multiplier, whose gain can be adjusted independently of the photon multiplier for the sideward channels, or the forward channels should be brought to the same distance as the sideward detectors. The latter would mean that we give up the one-degree angular resolution.

3.3.2. Characterisation of the polar nephelometer part of PHIPS-HALO

The PMMA fibre has a high numerical aperture (NA) of 0.7 (divergence angle about 23), which makes the coupling of the light from the parabolic mirrors to the fibre less sensitive to alignment errors. However, at the MAPMT side the high NA makes the coupling to the MAPMT more challenging and can lead to crosstalk between the MAPMT channels. The MAPMT channels are 0.8 mm wide with a 1 mm channel pitch, whereas the PMMA fibre diameter is 0.6 mm. Between the sensitive area of the photon multiplier and the fibre ends is a protecting borosilicate glass with a thickness of 1.5 mm that limits the optimum distance of the PMMA fibre ends. Therefore, we can expect a crosstalk due to the high NA of the PMMA fibres. To model this crosstalk optical simulations using the optical engineering software FRED (Photon Engineering, LLC, USA) were performed. The results from this simulation is shown in Fig. 3.19a. In an idealised case, where the optical fibres are placed directly on the protection borosilicate window of the MAPMT, a crosstalk of <2% can be expected. This crosstalk is the about same order as the crosstalk between the different MAPMT channels of 3% given by the manufacturer. If the distance of the fibre ends and the borosilicate window is increased, the expected crosstalk increases significantly.

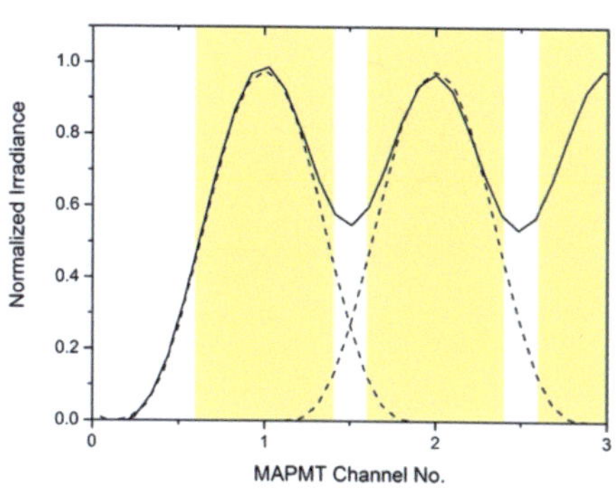

(a) Modelled crosstalk.

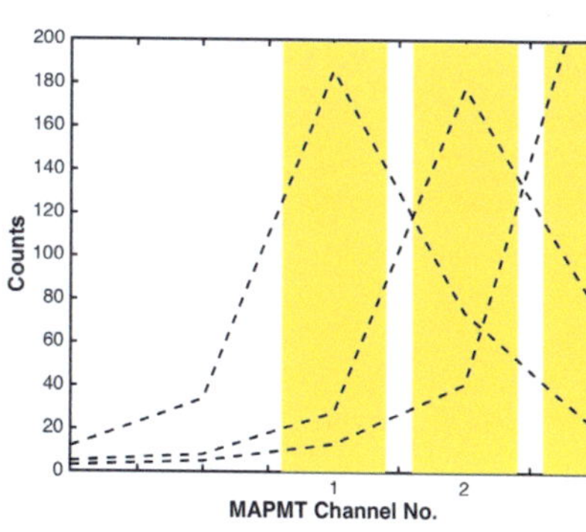

(b) Measured crosstalk.

Figure 3.19.: Characterisation of the MAPMT crosstalk. Fig. (a) adapted from Abdelmonem et al. (2016).

To validate the simulation results, I also performed laboratory measurements to define the crosstalk between the different channels. The laboratory setup consisted of an identical 532 nm Nd:YAG laser, an Ulbricht-Sphere (Labsphere Inc., USA) and an optical fibre terminated by fibre couplers. The laser light was guided to the Ulbricht-Sphere that is covered from inside with reflecting Spectralon material. The Ulbricht-Sphere diffuses the laser beam and reduces its intensity. One port in the Ulbricht-Sphere was equipped with an optical fibre adaptor so that the laser light could be conducted directly from the Ulbricht-Sphere to the PHIPS-HALO PMMA fibre using an optical fibre extension. Additional neutral density filters were used to reduce the laser light power so that the MAPMT signals were not saturated. I successively illuminated each of the 20 side- and backward detection optical fibres of the PHIPS-HALO instrument and measured, how the intensity was divided to the different MAPMT channels. A results from three channels is shown in Fig. 3.19b.

I found the crosstalk behaviour of the single PMMA fibres to be almost identical, and this crosstalk can be expressed with an averaged crosstalk matrix (table 3.2). The total crosstalk to the left side was 16.5% and 29% to the right side. This measured crosstalk was significantly higher than the modelled crosstalk, which indicates that in the real setup it is not possible to bring the PMMA fibre ends to the ideal position on the borosilicate window. The total crosstalk to the right side is somewhat higher than the crosstalk to the left side, which indicates that the position of the PMMA fibre array is shifted compared to the MAPMT. Therefore, it is crucial to perform this laboratory characterisation again, if the MAPMT is changed. I also observed that the crosstalk does not only concern the very next neighbouring channels but also affect the next four channels, although this influence is only the order of a few percentage.

Channel	Measured fraction of intensity [%]
n-4	0.9
n-3	1.5
n-2	3.5
n-1	10.6
n	54.8
n+1	21.1
n+2	4.9
n+3	2.1
n+4	1.0

Table 3.2.: An averaged crosstalk matrix from laboratory calibration.

The crosstalk can be considered significant and it is recommended to improve the optical design of the MAPMT coupler. It is not possible to correct the crosstalk with mathematical handling using a crosstalk matrix in the cases, where the scattering signal is low. This covers most of the small particles, whose angular scattering function have a minimum that is of the same order as the background. Therefore, I developed a method to correct the crosstalk in case of particle ensembles. This correction method is based on measurements with glass beads and will be discussed in the next section. Also to minimise the effect of crosstalk, the optical fibres were arranged in a respective order, i.e. the angular arrangement is mapped on the MAPMT, and the channel 106° was placed next to the trigger channel (channel 1 of the MAPMT) to avoid false triggers due to the crosstalk from the 170° channel that has a larger field of view and usually gets more scattered light than the sideward channels.

Crosstalk correction with glass bead measurements

The crosstalk increases the uncertainties to define characteristic features, like specular reflections, in the single particle angular scattering function. In case of angular scattering functions averaged over particle ensembles, these features are largely missing. Therefore, in this thesis I analyse only ensemble angular scattering functions that can be corrected using ensemble correction coefficients. I performed measurements with glass beads to define experimental correction coefficients for each channel. The experimental correction coefficients take into account the crosstalk as well as transmission efficiencies.

For the measurements I used two DRI-CAL (Duke Scientific) Polystyrene Divinylbenzene particle size standards: DC-20 and DC-50 with 20 µm and 50 µm nominal diameters, respectively, and a refractive index of 1.59. The DC-20 standard has a median diameter of 19.3 µm±0.7 µm and a standard deviation of 2.1 µm. The DC-50 standard has a median diameter of 49.7 µm±2.0 µm and a standard deviation of 3.4 µm. For both standards I calculated a theoretical scattering phase function using Mie theory based on the given size distribution and integrated over the solid angles of the detectors. The dry glass bead standards were dispersed into an air stream using a self-made disperser. This disperser works with pressurised air that is led through a metal tube to the bottom of a small air-tight bottle containing the standards. This aerosolises the standards and the overpressure in the bottle pushes the aerosol through a second metal tube at a higher position in the bottle. This fairly simple setup can easily be constructed and deployed in field campaigns for quick instrument check.

The theoretical (crosses) and the measured (coloured lines) raw angular scattering function of the glass bead ensembles are shown in the upper pan-

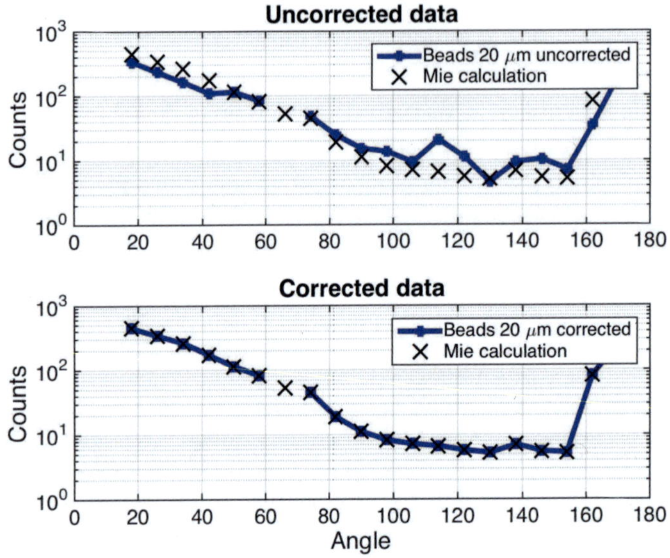

(a) Measurement with 20 µm glass standards.

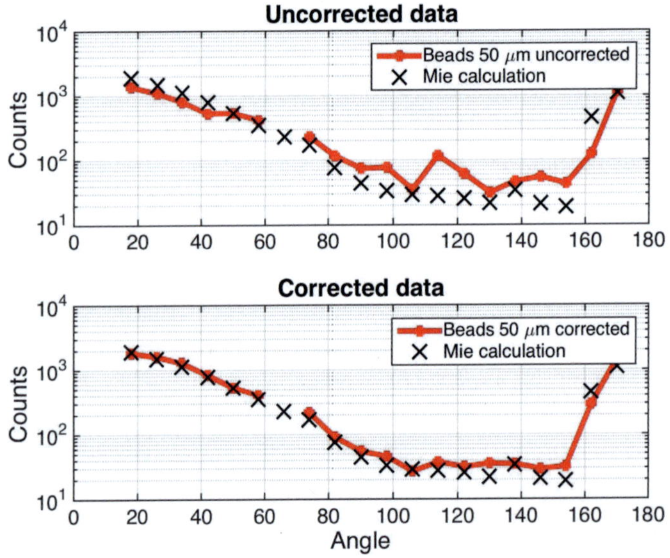

(b) Measurement with 50 µm glass standards.

Figure 3.20.: Glass bead measurements.

els of Figs. 3.20a and 3.20b. We see that the measured intensity at the forward (18-58°) and the backward (130-170°) channels is slightly lower than the expected intensity and on the contrary slightly higher in the sideward (74-122°) channels. The 66° PMMA fibre was not working during the measurements. This difference in the measured and theoretical intensities is the result of the crosstalk: the high intensity in the forward and backward scattering angles is transported to the sideward channels. Additionally, we see local maxima at 114° and in the 146° to 154° area in both cases. The maxima at 114° is due to the relocation of the 170° PMMA fibre to the neighbouring MAPMT channel. The other crosstalk feature in the 146° to 154° angular range is close to the rainbow angle that starts around 160° in case of glass spheres.

I corrected the measured intensities for the 20 µm glass beads to match the theoretical intensities. The 50° angle was set as reference and the other channels were corrected respectively. The correction factors for the channels from 18-170° are shown in table 3.3. The lower panel in Fig. 3.20a shows the crosstalk corrected measurements agreeing with the theoretical values. The same correction coefficients were then applied to correct the crosstalk of the 50 µm bead measurements (lower panel in Fig. 3.20b). We see a good agreement between the measurements of 50 µm glass beads that were corrected with the crosstalk correction factors for 20 µm glass beads and the theoretical calculations. This validates the crosstalk correction factors. The factors in table 3.3 are used in the data inversion to correct the averaged angular scattering function measurements.

18°	26°	34°	42°	50°	58°	74°	82°	90°	98°	106°	114°	122°	130°	138°	146°	154°	162°	170°
1.33	1.47	1.60	1.58	1	0.98	0.94	0.78	0.74	0.60	0.76	0.32	0.51	1.13	0.74	0.53	0.73	2.37	1.29

Table 3.3.: The crosstalk correction factors for the PHIPS-HALO angles 18-170°. The angle 66° was not in use and is excluded from the table. The reference angle was chosen to be 50° and, therefore, the correction factor is 1.

Inversion of the raw polar nephelometer data

In the polar nephelometer data analysis only the angles from 18-170° are included. The first 10 channels are not in use, as it was shown in the modelling study that the first 10 channels do net get enough intensity due to the scattering geometry. The raw data from the polar nephelometer is corrected using the following procedure:

1. **Background correction.** The PHIPS-HALO electronics, if not busy, send every $0.7\,\mathrm{s}$ a forced trigger and the force-triggered scattering functions can be used as a measure of the background intensity. The background corrected intensity is defined as

$$I_n = I_{0,n} - \bar{I}_{bg,n}, \tag{3.13}$$

where I_n is the background corrected raw signal in channel n, $I_{0,n}$ the raw signal in channel n and $\bar{I}_{bg,n}$ the mean background intensity over a time period in channel n. Usually, I use an averaging time of $10\,\mathrm{min}$ to get the mean background intensity from the forced trigger data. Only angular scattering functions that fulfil the condition $I_n > \bar{I}_{bg,n} + 3\sigma$ (σ is the standard deviation of the background intensity) are considered to have high enough intensity that they can be distinguished from the background.

2. **Saturation check.** The dynamic range of the MAPMT is limited in the analog-to-digital conversion to 11 bits, i.e. a maximum of 2048 counts. Individual scattering functions that show this maximum intensity in two or more channels are considered as saturated angular scattering functions and are excluded from the further analysis.

3. **Averaging.** At this point only angular scattering functions that are inside the dynamic range of the instrument and that are distinguishable from the background are left. From these individual angular scattering functions the mean raw scattering function during a time of interest is calculated.

4. **Crosstalk correction.** The crosstalk correction factors defined above are implemented.

5. **Smoothing.** The crosstalk correction may not remove all the crosstalk features. In the averaged angular scattering function, these residual features can be removed using a filter on the data. We used a moving average filter using a span of 4.

6. **Normalisation.** The resulting averaged angular scattering function is normalised so that its integral equals 1.

3.3.3. Defining the large-scale complexity from PHIPS-HALO images

The PHIPS-HALO instrument takes images with a maximum optical resolution of 2.35 µm (0.7 µm pixel resolution). Although this resolution is not enough to resolve the fine structure of ice particles, it is enough to give information of the large-scale complexity, like aggregation. Aggregation can potentially lead to similar angular light scattering properties than that of complex single crystals. Therefore, for modelling purposes the information of the aggregate versus single habit fractions is essential. The division between single habits and aggregates can be performed by manually classifying the PHIPS-HALO images. I performed a manual classification to a subset of $\sim$10 000 PHIPS-HALO images from the ACRIDICON-CHUVA campaign. However, for analysing larger datasets it is advised to use automated algorithms.

Schmitt and Heymsfield (2014) introduced a power law function to define a dimensionless parameter for representing the crystal complexity C'

$$C' = \frac{A}{\sqrt{ar}P^2}, \tag{3.14}$$

where A is the particle projected area, ar is particle area ratio (defined as the ratio of the particle projected area to the area of the smallest circle that will cover a two-dimensional image of the particle), and P is the perimeter of the image of the particle. These values can be derived from the PHIPS-HALO images using the above mentioned image analysis software. Equation 3.14 can be scaled as following

$$C = 10 \times (0.1 - C') \tag{3.15}$$

to get a complexity parameter C that is more intuitive in a way that increasing complexity is represented by higher values. Also the scale of the complexity is more reasonable; from 0 to 1. I performed the complexity analysis for the set of hand-classified images using Eqs. 3.14 and 3.15 to validate this method for the first time for the PHIPS-HALO instrument. This validation is shown in Sec. 6.

3.4. The measurement platforms

This thesis reports optical measurements of cloud and aerosol particles conducted on five different platforms. The aerosol optical properties were studied in aerosol chamber studies at the AIDA cloud simulation chamber and at the CERN CLOUD chamber. Ice crystal complexity and its link to ice crystal radiative properties were first investigated in cloud chamber studies with simulated cirrus and convective clouds at temperatures between -30°C to -60°C. Then, the knowledge gained from these cloud simulation studies was applied in field measurements onboard the German HALO aircraft, the Polar-6 aircraft as well as NASA WB-57 high altitude aircraft. The different field measurements targeted cirrus clouds in the tropics (ACRIDICON-CHUVA campaign) and in the mid-latitudes (ML-CIRRUS and MACPEX campaigns) as well as mixed-phase clouds in the arctic (RACEPAC campaign).

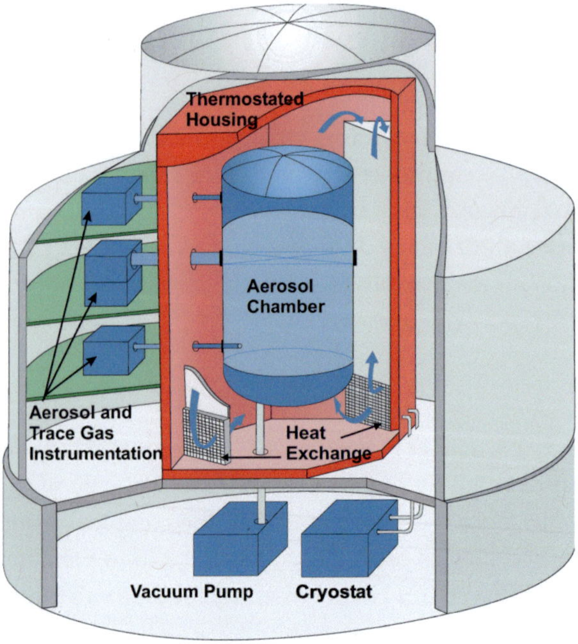

Figure 3.21.: A schematic image of the AIDA chamber.

3.4.1. The AIDA cloud simulation chamber

The Aerosol Interaction and Dynamics in the Atmosphere (AIDA) chamber of Karlsruhe Institute of Technology consists of a large aluminium (84 m^3) vessel that is located inside a thermal housing (Fig. 3.21). The large size of the chamber allows a long residence time of aerosol particles over hours or days, and a mixing fan located in the bottom part of the chamber maintains homogeneous conditions inside the chamber. This makes the AIDA chamber well suited for studying aerosol properties over longer time periods. The first investigations on aerosol optics in AIDA were done with carbonaceous aerosols (Schnaiter et al., 2003, 2005, 2006) and more recent studies have focused on the optical properties of dust aerosols (Linke et al., 2006; Wagner et al., 2012; Järvinen et al., 2016a).
Furthermore, the AIDA chamber can be cooled down to 183 K, which makes it suitable for studying ice microphysics in pure ice clouds (Schnaiter et al., 2012, 2016), in persistent mixed-phase clouds (Vochezer et al., 2016) and in convective systems (Järvinen et al., 2016c). To form liquid and ice clouds, supersaturated conditions inside the chamber are reached by expansion cooling; the chamber is evacuated from atmospheric pressure down to 600-800 hPa depending on the pumping speed and the required amount of cooling. The typical cooling rates that can be achieved at the beginning of the expansion range from $-1\,K\,min^{-1}$ to a maximum of $-2.5\,K\,min^{-1}$, which roughly corresponds updraft speeds from $2\,m\,s^{-1}$ to $7\,m\,s^{-1}$; values that are typical for mid-latitude convection over USA (Giangrande et al., 2013).

The experiments at AIDA are started by evacuating the chamber to a pressure below 1 hPa followed by flushing and cleaning cycles with particle free synthetic air, so that typical background aerosol concentrations are below 0.1 cm^{-3}. After that, a defined amount of pure water vapour is added to the evacuated chamber which is thereafter filled with dry synthetic air to obtain the desired relative humidity (RH). This procedure allows the experiments to be conducted in a wide range of RH conditions. The temperature of the chamber air is controlled by adjusting the temperature inside the thermal housing, i.e. the chamber walls, and it is constantly monitored by a set of horizontally and vertically aligned thermometers. The chamber pressure is monitored with a Baratron (MKS, Germany) pressure sensor and the RH inside the chamber is monitored by a fast high precision chilled mirror hygrometer (MBW, model 373) and with a tuneable diode laser setup (TDL, Fahey et al. (2014)).

Besides the basic chamber instrumentation, additional cloud probes can be installed underneath the chamber. These probes are first installed in pressure and temperature controlled canisters that are connected to vertical sampling inlets emerging from the chamber bottom. Since the sampling is vertical, we avoid deposition losses of the larger cloud particles. A more detailed description of the expansion techniques, as well as the chamber preparation and instrumentation can be found in Möhler et al. (2003), Wagner et al. (2009) and Wagner et al. (2011).

3.4.2. The CERN CLOUD cloud chamber

The CERN CLOUD[4] chamber (Duplissy et al., 2015) is installed at the T11 beam line in the east hall of the CERN Proton Synchrotron (Fig. 3.22a). The location of the CLOUD experiment was chosen to study the effects of electrification and galactic cosmic rays to atmospheric chemistry and new particle formation (e.g. Kirkby et al., 2011). In 2013, the CLOUD chamber was upgraded to work as an expansion chamber. This allowed series of cloud and aqueous chemistry experiments that were conducted in two campaigns between 2013 and 2014 (CLOUD8 and CLOUD9). This section describes the chamber operation during those campaigns.

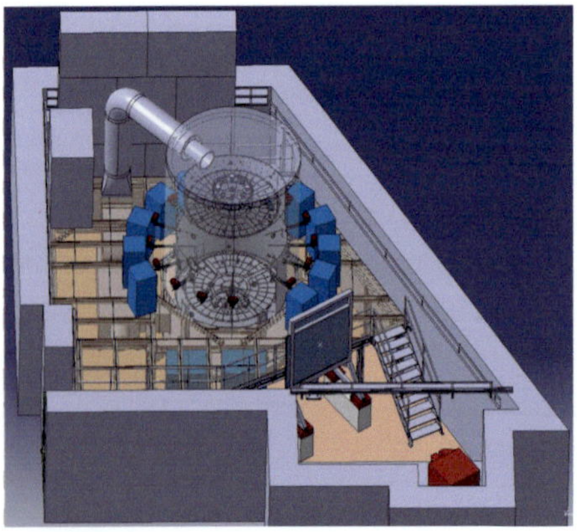

(a) Position of the CERN CLOUD chamber in the T11 hall. From Duplissy et al. (2015).

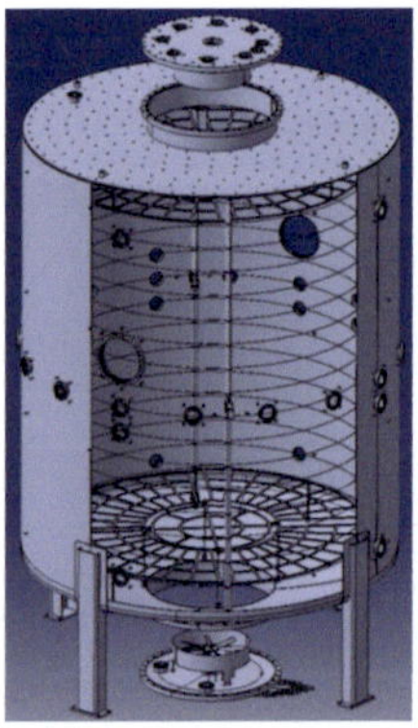

(b) A schematic image of the CERN CLOUD chamber.

Figure 3.22.: The CERN CLOUD chamber

[4] CLOUD is a acronym for Cosmics Leaving Outdoor Droplets

A schematic image of the CERN CLOUD chamber is shown in Fig. 3.22b. The chamber consists of a 26 m^3 stainless steel cylinder that is located inside a housing with thermal insulation. The stainless steel allows a high standard for chamber cleanliness and, additionally, the chamber can be flushed with high-pressure ultra-pure water and heated to 100°C to reduce contaminations. The temperature of the chamber can be varied from -60°C to 100°C and it is controlled by regulating the temperature of air flowing around the chamber and inside the thermal housing. The temperature inside the chamber is measured with a string of six thermocouples positioned horizontally at different distances between the wall and the chamber centre. For reporting the chamber temperature, I used the mean of the four inner thermocouples as a representative measure of the temperature in the chamber volume.

To initiate an expansion, the chamber volume is first pressurised to a maximum overpressure of 250 mbar. Adiabatic cooling is created by releasing the pressure through a series of valves. The speed of the expansion can be controlled, so that expansion rates up to 10 mbar s^{-1} can be achieved, leading to a maximum cooling rate of 22 K min^{-1}. This roughly corresponds to an updraft speeds of 50 m s^{-1}. Therefore, the experiments at CERN CLOUD chamber complement the dynamic range of the AIDA chamber and allow simulations of convective systems from low convection up to thunderstorms. In section 5.1, I report results from simulated thunderstorm convective systems conducted at the CERN CLOUD chamber.

In the course of static experiments at the CLOUD chamber, air from the chamber is constantly sampled; and this sampled air is replaced with humidified artificial air to maintain a constant pressure inside the chamber. The artificial air is created by evaporating liquid nitrogen and oxygen, and humidified by passing part of the flow through a Nafion humidifier to achieve a chosen constant RH. Alternatively, the air is passed through a heated line to the chamber, allowing the dew point of the incoming air to be higher than that inside the chamber. The water vapour is mixed with the main air stream before entering the chamber, which allows a homogeneous RH throughout the chamber volume after a mixing time of a few minutes (Voigtländer et al., 2012). The RH inside the chamber is measured with a chilled mirror dew point hygrometer (MBW, model 973). During CLOUD9 a tuneable diode laser (TDL system) was used to measure water vapour, in addition to the MBW.

3.4.3. Airborne measurements onboard HALO

HALO[5] is a German research aircraft based at the National Aeronautics and Space Research Centre of the Federal Republic of Germany (DLR[6]). The aircraft is a modified Gulfstream G550 business jet that was transformed into a research platform. The advantages of the HALO aircraft are the high cruising altitude of up to 15 km, a long range well above 10 000 km and a large payload. This makes HALO the only research aircraft in Europe capable of studying cirrus clouds in the upper troposphere and lower stratosphere (UT/LS).

The HALO can accommodate in-cabin instrumentation as well as underwing instrumentation, like cloud probes. The underwing stations were installed at the end of 2013 and certified at the beginning of 2014. The first scientific mission with the underwing instrumentation was the Mid-Latitude CIRRUS experiment (ML-CIRRUS) over Europe in spring 2014. The same year the HALO aircraft with the underwing instrumentation participated in the German-Brazilian ACRIDICON-CHUVA campaign in Brazil. In both campaigns, the SID-3 and PHIPS-HALO instruments were operational onboard HALO.

3.4.4. Airborne measurements onboard Polar-6

Arctic mixed-phase clouds over sea-ice covered areas were targeted in the Radiation-Aerosol-Cloud Experiment in the Arctic Circle (RACEPAC) campaign in April-May 2014 as a collaboration of German and French research institutes. The goal of the campaign was to study the microphysical and radiative properties of arctic mixed-phase clouds over the Beaufort sea to achieve a better understanding of the role of mixed-phase clouds in the arctic warming. The campaign was coordinated by the group of M. Wendisch, LIM, University of Leipzig.

Measurements were conducted with two aircrafts (Polar-5 and Polar-6) operated by the Alfred Wegener Institute (AWI) for Polar and Marine Research, Germany. The AWI Polar aircrafts are modernised Basler BT-67 planes. Updates have been made to allow operation at cold arctic regions. During RACEPAC, coordinated flights were made with the two Polar aircrafts. The Polar-5 was equipped with remote-sensing instrumentation, while the Polar-6 carried in-situ cloud, aerosol and trace gas instrumentation. The in-situ cloud instrumentation included the SID-3 and PHIPS-HALO instruments. While the PHIPS-

[5] HALO is an acronym for High Altitude and LOng range

[6] The acronym DLR comes from the German name Deutsches Zentrum für Luft- und Raumfahrt e.V.

HALO was still in its development phase, the SID-3 collected an unique data set of the small mixed-phase ice particles.

3.4.5. Airborne measurements onboard NASA WB-57

The NASA[7] WB-57 is a mid-wing, long-range aircraft that can reach an altitude of 18 km and has a range of 4000 km. The aircraft type was former used for intelligence flights but since 1970's they have been used for research purposes. Like the HALO aircraft, the WB-57 can carry several underwing instrument probes. The SID-3 was operational onboard the WB-57 during the Mid-latitude Airborne Cirrus Properties EXperiment (MACPEX, March/April 2011). The WB-57 was operated out of Ellington Field, TX, from where research flights were conducted to investigate the properties of mid-latitude cirrus clouds, the processes affecting these properties and their impact on radiation. Overall 14 research flights were conducted, during which the SID-3 collected a comprehensive data set of mid-latitude cirrus ice particles.

[7] The National Aeronautics and Space Administration

4. The near-backscattering depolarisation properties of complex dust particles and viscous secondary organic aerosol

In chapter 2 we learned that the backscattering depolarisation ratio can be used as a measure of particle asphericity, as for a perfectly isotropic, spherical particle, light that is scattered directly back toward its source, i.e., in a direction of 180° from that of the incident light, retains its initial polarisation state. Whether it results from deviations from spherical symmetry, from complexity, or from structural inhomogeneities, anisotropy in the optical properties leads to depolarisation. A measurement of the extent of depolarisation in the backscattered light probes the magnitude of the anisotropy and can be used to detect aspherical or anisotropic particles in a number of optical applications.

One specific optical application that relies on the backscattering depolarisation measurements is the polarisation lidar technique using several optical parameters to retrieve the microphysical properties of the target particles. Information of the particle backscattering depolarisation properties are used in lidar retrievals to distinguish between liquid, mixed-phase and ice clouds. Also layers of aspherical aerosol particles can be identified based on their non-zero depolarisation and low backscattering ratios. Recently, a lot of attention has been given to distinguish complex dust particles and to retrieve their distributions in the atmosphere.

Although the lidar method is highly promising, the basic problem remains: the microphysical properties need to be solved from a number of optical parameters through an inversion problem that is always ill-posed and relies on accurate solutions of the forward problem. In the case of complex particles, the forward problem cannot be solved analytically, but empirical or theoretical approaches need to be taken. In this section, I experimentally investigate the forward problem in two complex aerosol systems: the depolarisation ratio of complex dust particles and the detection of the viscous phase state of SOA using depolarisation measurements. The results and discussion in this section are modified from my first authored publications Järvinen et al. (2016a) and Järvinen et al. (2016b).

4.1. Laboratory investigations of the near-backscattering depolarisation ratio of complex dust particles

Coarse-mode dust particles are known to have a high linear depolarisation ratio compared to fine-mode aerosol (biomass burning smoke, urban haze) and, therefore, they can be distinguished from fine-mode aerosol particles (e.g. Sugimoto et al., 2003; Shimizu et al., 2004; Sugimoto and Lee, 2006; Omar et al., 2009; Tesche et al., 2009; Ansmann et al., 2011; Groß et al., 2011; Ansmann et al., 2012; Burton et al., 2012; Müller et al., 2013). On the other hand, the Aerosol Robotic Network (AERONET) sun/sky photometer observations have shown that a significant fraction of dust-related optical depth is linked with fine-mode dust (Mamouri and Ansmann, 2014). These new observations have led to a need to develop new techniques to determine the fraction of fine-mode dust to coarse-mode dust (e.g., Mamouri and Ansmann, 2014).

To retrieve the fractions of coarse- and fine-mode dust, assumptions of the scattering properties of these complex particles are to be made. These assumptions are either based on empirically gathered data (e.g. Ansmann et al., 2012) or on modelled single scattering properties of dust particles (e.g. Dubovik et al., 2006; Chaikovsky et al., 2012). However, due to the computational limitations, most of the scattering models are restricted to model particles with simple geometries and uniform refractive indices, even though real dust particles are complex, irregular and inhomogeneous (Nousiainen, 2009). As the computation power is increasing, new, more realistic models using complex geometries or inhomogeneous particles (e.g., Ishimoto et al., 2010; Gasteiger et al., 2011; Lindqvist et al., 2011; Liu et al., 2013; Kemppinen et al., 2015a,b) could significantly improve the representation of dust particle scattering properties in lidar retrievals. However, these models still involve significant approximations and, therefore, validation of them require laboratory measurements of well-characterised dust samples.

There are only few laboratory studies available studying the backscattering or near-backscattering depolarisation properties of natural mineral dust particles. Muñoz et al. (2007) measured the whole scattering matrix, including the depolarisation ratio, of Saharan dust particles. The angular range they covered was from 4° to 174°, but the setup was later improved to cover angles up to 177° (Muñoz et al., 2010). The measured dust sample contained a mixture of particles with sizes ranging from a few microns to several hundred microns, leading the dust scattering properties to be dominated by the coarse-mode dust particles. The first step towards size-segregated measurements was taken by Sakai et al. (2010). The authors were able to experimentally show that the dust near-backscattering (angular range from 178.8° to 179.6°) depolarisation properties are size-dependent by using an impactor to remove coarse-mode particles from the dust sample. The authors found out that the fine-mode dominated dust sample showed a significantly lower depolarisation ratio (around 15%) than the coarse-mode dominated dust sample (δ_L around 40%). This laboratory study since served as an important basis for lidar studies (e.g. Ansmann et al., 2012; Mamouri and Ansmann, 2014; Groß et al., 2015). In a more recent study, Miffre et al. (2016) measured at the exact backscattering angle (180±0.2°) depolarisation ratio of two laboratory produced Arizona Test Dust size distributions. The authors performed the measurements at two wavelengths (355 and 532 nm), corresponding to common wavelengths used in lidar applications.

Still, significant experimental and theoretical work is needed to evaluate the scattering properties of natural dust particles. The experimental work is usually limited to uncertainties in aerosol properties, such as composition and size distribution. In this section, I address the latter problem and present size-segregated laboratory measurements of near-backscattering linear and circular depolarisation ratios for mineral dust particles of natural and industrial origin. I compare the measurements with results from three different scattering models using stereogrammetric dust particles (Lindqvist et al., 2014), inhomogeneous Voronoi tessellation particles (Kemppinen et al., 2015a) and spheroidal particles (Mishchenko et al., 1996).

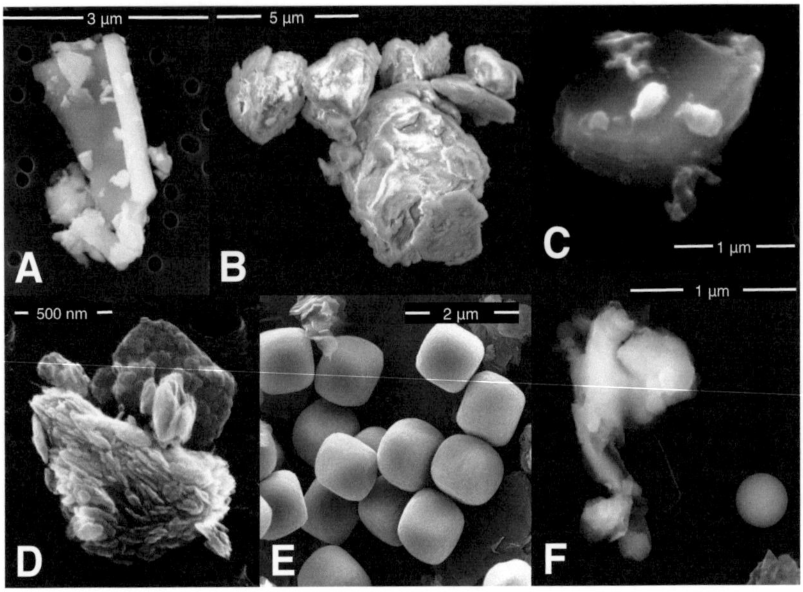

Figure 4.1.: Scanning electron microscope (SEM) images of dust particles used in the experiments. Particle (A) is a volcanic ash particle from Eyjafjallajkull volcano eruption, (B) a desert dust particle from Israel, (C) a soil dust particle from Karlsruhe, (D) an Illite particle, (E) a cubic hematite particles and (F) an ATD particle coated with secondary organic aerosol. A homogeneously nucleated secondary organic aerosol particle is seen next to the coated ATD particle.

4.1.1. Experimental setup

The experimental results cover 10 years of dust depolarisation measurements at the AIDA chamber. Dust samples of different origin have been used as cloud condense nuclei or ice nuclei in cloud droplet formation or freezing experiments in several AIDA campaigns. Before the dust aerosol is activated in an expansion, the dust is characterised for its size distribution and sometimes for its optical properties. Here, I have analysed the AIDA experiments, where both depolarisation and dust size distribution measurements were conducted simultaneously before the activation. To advance the existing dataset with measurements of circular depolarisation ratio, I performed a dedicated depolarisation campaign PÖLY01[1] in spring 2015. In the following two chapters I describe

[1] PÖLY is a Finnish word for dust

the aerosol dispersion and characterisation methods that are used in all AIDA experiments and go through the experimental procedure.

Aerosol preparation and characterisation

The dust samples (Fig. 4.1) were introduced to the chamber using a rotating brush generator (RBG 100, Pallas). After the dust injection, the settling and inertial losses limit the upper particle diameter to 10 µm. Furthermore, in some experiments larger dust particles were removed using one or two cyclones with different cutoffs: the first cyclone impactor has a cutoff of 50 % between about 1 µm and 2 µm and the second has a cutoff of 50 % between about 2 µm and 4 µm. During the injection process the particle number concentration was continuously monitored using a condensation particle counter with 10 nm cutoff (TSI, CPC3010).

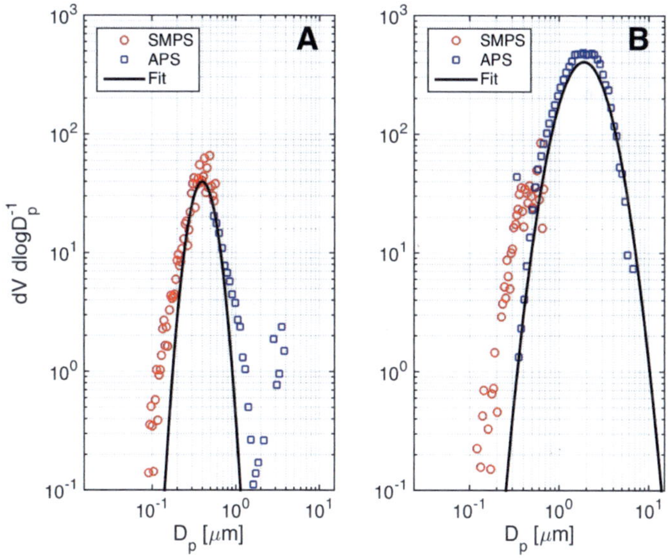

Figure 4.2.: Examples of the mono-modal size distributions of dust particles generated for the experiments. Size distribution A is from an experiment with Illite particles with mean volume equivalent diameter of 0.4 µm and mode width of 1.4 and B from experiment with Argentinian soil dust with mean volume diameter of 2.4 µm and mode width of 1.6.

The size distribution of the dust aerosol in the chamber was derived from the combination of aerodynamic particle sizer (APS) and scanning mobility particle sizer (SMPS) measurements. The SMPS measures the electrical mobility diameter of aerosol particles in the size range of 14–820 nm. In the APS instrument, aerosol particles in a size range of 0.5–20 µm are measured based on the aerodynamic diameter. The data from the two instruments was combined following the method by Möhler et al. (2008), where both the mobility equivalent and the aerodynamic diameters are converted into the volume equivalent diameter. Then, a lognormal fit is done for the combined aerosol size distribution and the volume equivalent lognormal parameters were defined for each dust sample.

Figure 4.2 shows two size distributions measured for Illite particles (Fig. 4.1D) and for Argentinian soil dust particles. Both size distributions are monomodal (although few larger Illite particles are remaining in panel A) and have relatively narrow geometric standard deviations (σ) of 1.4 and 1.6 for the Illite particles and Argentinian soil dust, respectively. In general, σ of the produced size distributions varied between 1.2 and 2, although the majority of the values were below 1.6. Some size dependency was observed in σ, as σ decreases towards smaller size distributions.

Additionally, in some experiments the dust particles were coated with either SOA or with SA. A detailed description of the procedure for in-situ SOA coating can be found in Möhler et al. (2008) and for the SA coating in Möhler et al. (2005). Fig. 4.1 panel F shows an ESEM image of a SOA coated Arizona test dust (ATD) particle. In this case some of the organic vapour nucleated homogeneously to form externally mixed, spherical SOA particles as one is seen next to the ATD particle in Fig. 4.1. Since the presence of spherical particles will affect the ensemble depolarisation properties, cases with external mixtures were excluded and only internal mixtures were considered in this thesis.

The experimental procedure can be summarised as following, according to Fig. 4.3.

Experimental procedure

1. Before injecting the dust particles, the aerosol chamber was cleaned and humidified using a procedure described in section 3.4.1.

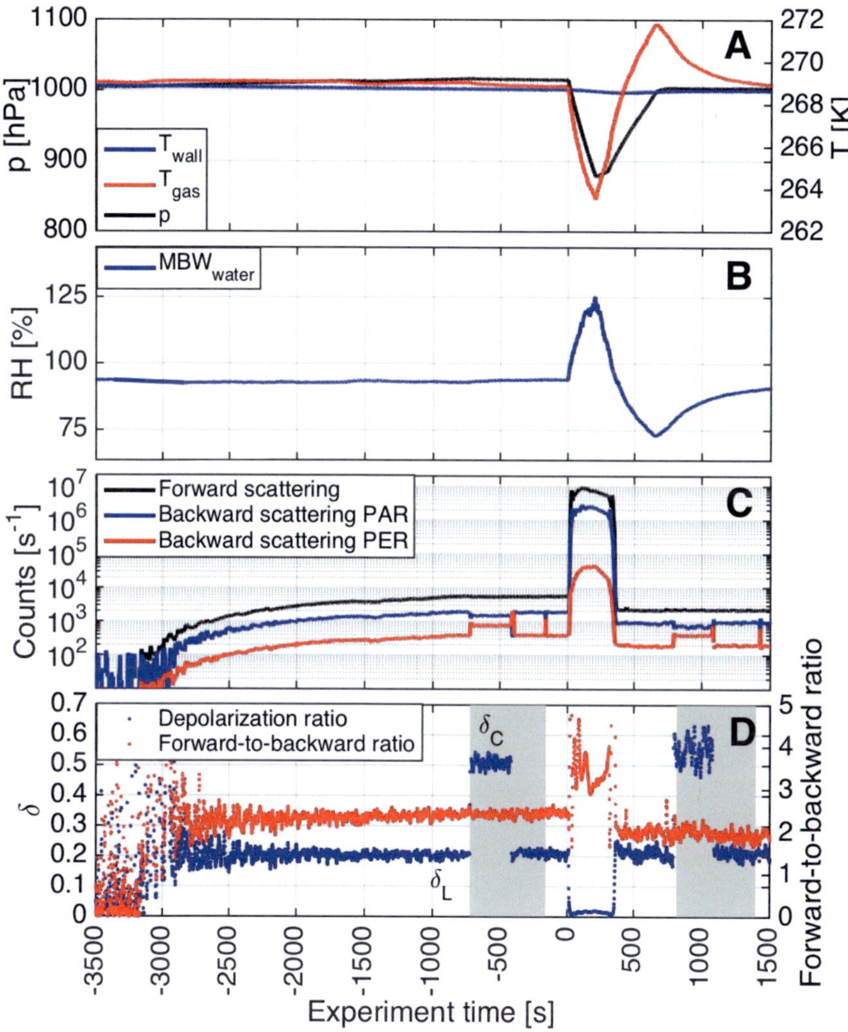

Figure 4.3.: Example of an experiment run with Argentinian soil dust. Panel A shows the gas and wall temperatures as well as the chamber pressure. Panel B shows total RH (condensed plus gas phase water) inside the chamber, panel C the near-forward and near-backward scattering signals, and panel D the depolarisation ratio (blue) and the forward-to-backward scattering ratio (red). An expansion was started at experiment time 0 s to activate the dust particles to form cloud droplets.

2. The dust particles were injected into the clean chamber using the procedure described above. I used only one dust sample per experiment, and before changing to another dust sample, the chamber was cleaned. In Fig. 4.3 the injection of the dust is started at experiment time −3100 s. The increase in the dust number concentration is seen by the increase of the scattering signals in panel C. The injection of the dust aerosol was continued until the scattering signals were significantly higher (at least twice as high) than the background signal. This resulted in aerosol particle concentrations of typically around 300–700 cm^{-3} depending on the diameter of the particles.

3. The mixing fan inside the chamber enabled the aerosol to reach well-mixed conditions within minutes after the aerosol addition was stopped. The stability of the aerosol population in terms of the size, shape, and composition distribution is seen by the almost constant scattering signals in Fig. 4.3 panel C and in the constant depolarisation signal and forward-to-backward scattering ratio in panel D between experiment time −1500 s and 0 s.

4. I measured the depolarisation ratio of the stable aerosol population simultaneously with the size distribution measurements. These measurements are highlighted in grey in panel D. After having the SIMONE-Junior instrument available at the chamber, I measured also the circular depolarisation ratio during these periods. The change to circular incident polarisation state is seen in the change of the scattering signals in Fig. 4.3 panel C and as jumps in the depolarisation ratio in panel D between experiment time −685 s and −417 s.

4.1.2. Depolarisation ratio as a function of size

In this work, I analysed over 200 dust samples and linked their depolarisation properties to their measured size distribution. Overall, the dust linear depolarisation ratio, δ_L, showed a median value of 0.23 and a statistical variation between 0.03 and 0.36 (5 and 95 percentile). The large variability in δ_L can be explained either with the differences in dust particle size, shape and composition. I studied the effect of size by producing well-defined mono-modal dust particle size distributions with different volume median diameters (VMD). The effects of composition and shape were investigated by using dust samples from different source regions, assuming that differences in the dust particle shape and composition are depending on their source. The results in terms of the functional dependency of the linear depolarisation ratio with respect to VMD are shown in Fig. 4.4. The measurement results can be summarised with three conclusions:

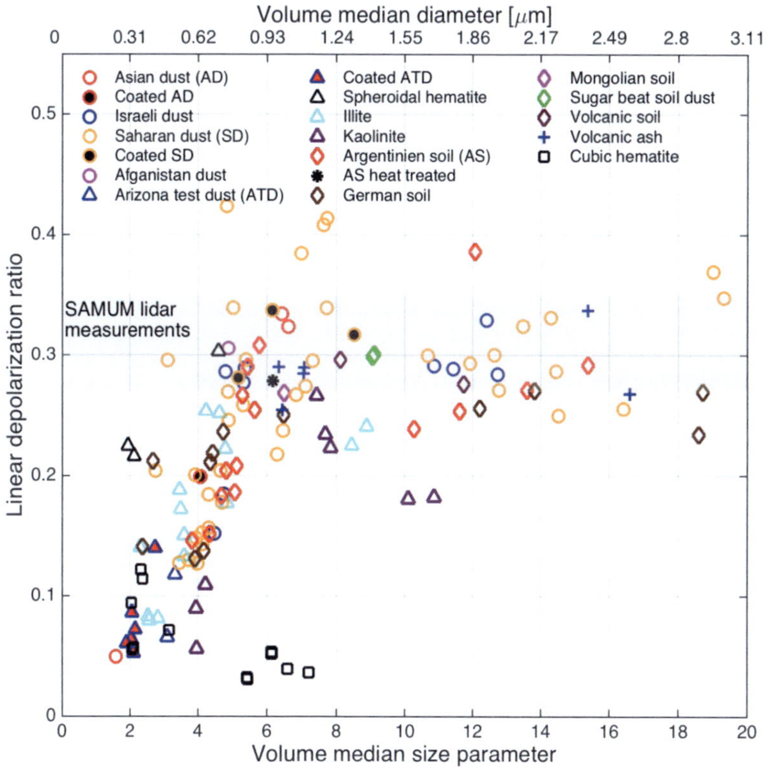

Figure 4.4.: Linear depolarisation ratios of desert dust samples (square symbols), commercially available dust samples (triangles), soil dust samples (diamonds), volcanic ash (blue crosses) and cubic hematite (black squares) as function of particle mean diameters. Some of the dust samples were coated with sulphuric acid or with secondary organic material (filled symbols). The grey filled area represent the range of linear depolarisation ratios (0.27-0.35) measured for atmospheric Saharan dust layers during the SAMUM field campaign (Freudenthaler et al., 2009).

i) The linear depolarisation ratio of complex dust particles depends on the dust particle size.

ii) The same trend in the depolarisation ratio is seen for all dust particles independent of their source region, except in the case of artificial cubic hematite.

iii) Coating of the dust particles with sulphuric acid or secondary organics in atmospheric ageing process does not influence their depolarisation properties.

Although point i) is predicted from theoretical studies (Nousiainen, 2009; Gasteiger et al., 2011; Nousiainen et al., 2012), the results show a size-dependence on an experimental basis. Furthermore, Fig. 4.4 shows a clear separation in the depolarisation properties of fine- and coarse-mode dust particles, as previously indicated by Sakai et al. (2010). Over the fine-mode size range δ_L strongly increases from 0.05 (measured for 0.25 µm particles) to maximum values between 0.27 and 0.41 for median diameters around 1 µm. The coarse-mode δ_L values are generally higher than the fine-mode values. Over the coarse-mode size range δ_L varies between 0.18-0.41 with a mean value of 0.27 and stays rather constant or even slightly decreases. The fine-mode depolarisation values agree well with the study of Sakai et al. (2010), but the coarse-mode depolarisation values are on average somewhat lower compared to the reported mean value of 0.39, although several individual depolarisation values were found to be comparable.

Points ii) and iii) are new and somewhat surprising. A possible explanation for observation ii) is a similar and high complexity in terms of shape, surface roughness and composition between the different dust samples that results in similar depolarisation properties of the airborne particle ensembles. This trend is more clear for fine-mode particles than for the coarse-mode particles, as smaller particles can be less sensitive to small-scale details in the shape (Nousiainen, 2009). For the same reason, the variation in the depolarisation properties of the different dust samples is more profound for the coarse mode particles. For example, Saharan dust particles have a relatively significant span in δ_L of 0.2 at VMD=1 µm. Also, in the coarse-mode we observe greater differences in the depolarisation properties between the different dust samples than in the fine-mode. The largest differences in δ_L are seen between natural dust samples (marked as circles and diamonds in Fig. 4.4) and commercially available dust samples (marked as triangles and squares in Fig. 4.4). The commercially available dust samples show somewhat more profound decrease in δ_L towards larger diameters, while δ_L of natural dust samples seems to remain more constant.

The only exception to point ii) was cubic hematite, as their depolarisation properties show a clear separation to the complex and irregular samples. The measured δ_L for cubic hematite were consistently below 0.15. The low δ_L can be explained with the shape of the hematite particles, as Kemppinen et al. (2015a) showed that a cube, even with a more dust-like refractive index, has lower linear depolarisation ratios than most of the more irregularly shaped particles with the same refractive index, and that this difference becomes smaller if the surface of the cubic particle is roughened.

Understanding the effects of dust particle size, shape or composition to their depolarisation properties only on the basis of measurements is not straightforward. It has to be kept in mind that the presented interpretation of observations i) and ii) does not necessary mean that in the case of complex and irregular particles only their size defines their depolarisation properties, as the other particle properties, such as shape (or degree of complexity) and refractive index, can be size dependent.

4.1.3. Comparison of measurements and simulations

I compared the measurement results to simulation results using three distinct particle models: stereogrammetric particles, tessellation particles and spheroids (Fig. 4.5). The shape of the stereogrammetric particles is based on images of real dust particles. For this work four particles with geometries determined by stereogrammetry were used. The particle names, with brief descriptions of their compositions, are Cal (calcite), Dol (dolomite), Agg (aggregate of several minerals, quartz being the most abundant), and Sil (silicate, mostly chrysotile). The particles thus vary both in shape and in composition. Because stereogrammetry may not present the particle surface roughness accurately, artificial surface roughness was introduced by following the method of Kemppinen et al. (2015b).

In addition to using realistic particle geometries, it is also interesting to know how particle inhomogeneity and internal structure affects linear depolarization ratio. A method was developed for generating realistically shaped inhomogeneous particles by Kemppinen et al. (2015a). In this study, five different internal structure cases were generated. First, a particle with only mutually similar clay minerals (Orig). Second, the first case, but with additional hematite (a strongly refracting and absorbing mineral) nodes in the particle (Iron). Third, the first case, but with internal pores (Nodes1). Fourth, the first case, with both the hematite nodes and the internal pores (Nodes2). Fifth, finally, the first case, but with coating material replaced by a hematite-rich clay mixture (Nodes3). The base mineral compositions are described in detail in (Kemppinen et al., 2015a), and are based on real Asian dust particle compositions. The last particle model was the spheroidal model without any inhomogenities, which was used as a comparison for the more realistic dust particle models.

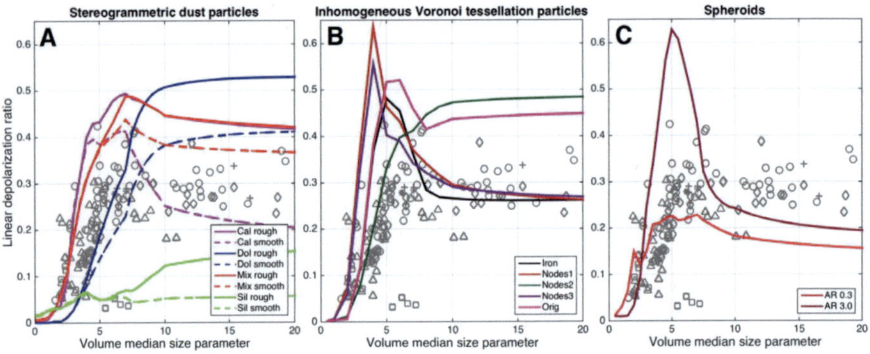

Figure 4.5.: Comparison of modelling results and laboratory measurements. Panel A shows the modelling results for stereogrammetic dust particles with roughened surfaces (solid lines) and with smooth surfaces (dashed line). Panel B shows the modelling results for inhomogeneous Voronoi tessellation particles (solid lines). Panel C shows the modelling results for spheroidal particles using the T-matrix approximation. The calculations were made for two aspect ratios (AR): for a prolate particle with AR=0.3 and for an oblate particle with AR=3.0.

Generally, every scattering model (with exemption of stereogrammetic Sil) showed a similar size-dependence of δ_L: a steep increase in δ_L was seen up to one micrometer, whereafter the δ_L decreased to a constant value or stayed unchanged. The modelling results were found to be highly sensitive to assumptions of the particle size distribution. Especially, the assumed mode width influenced the results significantly at size parameters between 1-5, where the δ_L increases from zero to close to maximum value. This highlights the importance of accurate size distribution measurements, as a small error in the size determination can lead to large discrepancies between measurements and models.

Within the different scattering models, large differences were observed in the depolarisation properties of different model particles. In the stereogrammetic particles model (Fig. 4.5, panel A) the best particle types to represent the measurements were found to be Cal or an external mixture of different particle types (Mix). Dol predicted too high values for δ_L, whereas Sil showed almost constant and low δ_L values that agreed with the measurements for cubic hematite. The calculations for stereogrammetic particles were done for smooth and rough particle versions to study the influence of roughness in the model. The ESEM images of our dust samples (Fig. 4.1) clearly indicated that surface roughness is a common feature in real dust particles. However, the calculations for smooth particles matched the measurement better than the calculations for their roughened counterparts. Adding roughness to the model particles increases their δ_L values. In the case of the Cal and Mix -particles the increase was $\sim$0.2, leading to δ_L values above 0.4.

In the model with Voronoi tessellation particles (Fig. 4.5, panel B), the modelling results of different particle types show little variation for sub-micrometer diameters, but a larger span for larger diameters. The measurements compare well with inhomogeneous iron-coated particles and particles with iron nodes. The fact that the tessellation particles with iron content matched the measurement data significantly better than the particles without iron content may indicate that iron content is a common feature in natural dust particles. However, as shown by Kemppinen et al. (2015c), a good match in an individual scattering quantity does not necessarily imply a match in all microphysical properties, especially if there are more than one unknown property, such as shape, composition or surface roughness. I also studied the homogeneous versions of the same particles, but the model results clearly did not fit the measurement results, showing significantly higher values than the inhomogeneous particles.

The last model calculations were done for an ensemble of randomly oriented spheroids. Although only two axis ratios are considered, they highlight the very wide range of values that the spheroidal shapes can provide by changing the model parameters. However, at median size parameters $>$10, the calculations do not seem to match the measured values of dust at the used refractive index. A better agreement between the spheroidal model and the measurements could be achieved by tuning the refractive index.

4.1.4. On the symmetry assumption in the case of complex dust particles

In case of light scattering by particle ensembles in random orientation, the scattering matrix becomes diagonal, if the following symmetry conditions are fulfilled: (i) the particles are either plane symmetrical (as in the case of spheroids) or (ii) a system is large enough that each particle has a mirror particle in the described system. In this case a general relationship between the backscattering (180°) linear (δ_L) and circular depolarisation (δ_C) ratio is valid, and this relationship can be expressed as (Mishchenko and Hovenier, 1995b)

$$\delta_C = 2\delta_L/(1 - \delta_L). \tag{4.1}$$

Therefore, simultaneous measurements of linear and circular depolarisation ratios can be used to experimentally study the symmetry of a system. Polarisation lidar techniques have been used to investigate the symmetry assumption in cirrus or liquid clouds (Del Guasta et al., 2006; Hayman et al., 2012), but the symmetry in aerosol systems is rarely studied.

I used the SIMONE-Junior instrument, capable of alternating between linear and circular laser incident polarisation, to measure the dust near-backscattering linear and circular depolarisation ratios in the chamber experiments. Although the measurements were not conducted simultaneously, the aerosol population remained stable during the measurement phase (~10 minutes), and therefore, it can be assumed that the measurements represented the same aerosol sample. Figure 4.6 shows the measurements together with the theoretical equation (Eq. 4.1) for symmetry. The measurements follow the theory without exception, showing a symmetry in the system. We can assume that due to the mixing in the chamber, the dust particles were randomly oriented. In the case of randomly oriented particles, the symmetry assumption have been found to be valid in cirrus clouds (Del Guasta et al., 2006; Hayman et al., 2012) and now also in dust aerosol. However, it is not obvious, why this assumption should be valid, as we cannot expect the ice or dust particles to have plane symmetry. The particle ensemble, therefore, needs to consist of mirror particles. In the measured particle ensembles, the high degree of complexity of individual particles might increase the probability of finding mirror particles.

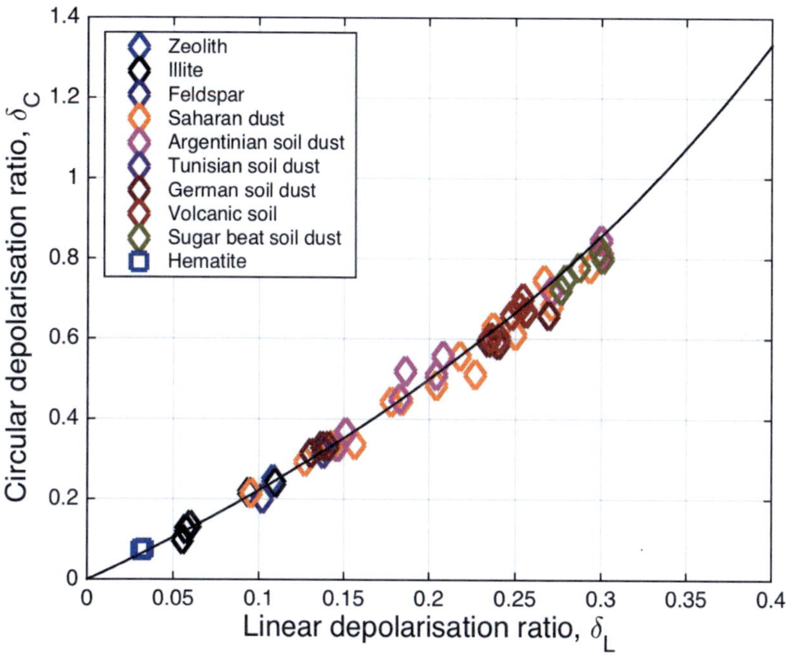

Figure 4.6.: Relation between measured linear and circular depolarisation ratio for dust samples. The black curve shows the theoretical relationships of the linear, δ_L, and circular, δ_C, backscattering depolarisation ratios for non-spherical particles in random orientation (Mishchenko and Hovenier, 1995b).

4.2. Light scattering as a tool to detect the viscous phase state of SOA

Recently, considerable attention has been given to water uptake and viscous properties of secondary organic aerosol (SOA). It has been found that SOA particles can exist in the atmosphere in an amorphous semi-solid or solid state (Virtanen et al., 2010; Koop et al., 2011; Renbaum-Wolff et al., 2013; Pajunoja et al., 2014). The existence of these states has several atmospheric implications. Molecular diffusion in the condensed phase affects the gas uptake by viscous SOA particles and can, therefore, alter the SOA particle lifetime in the atmosphere (Shiraiwa et al., 2011). In addition, water uptake is inhibited in the viscous particles (Riipinen et al., 2012), limiting the SOA growth under conditions, in which SOA would typically grow hygroscopically (Swietlicki et al., 2008; Pajunoja et al., 2015). This, in turn, influences the aerosol direct effect on radiative forcing. More recent studies have shown that the viscous SOA particles or their proxies can act as ice nuclei (Murray et al., 2010; Wagner et al., 2012; Wang et al., 2012; Wilson et al., 2012; Schill et al., 2014), thus influencing cloud cover, cloud optical properties, and precipitation.

In order to understand the climatological influence of the viscous state of SOA particles, detailed knowledge of the temperature and relative humidity (RH) ranges in which SOA can persist in a solid or semi-solid state is needed. Several experimental methods have been developed to measure the transition temperature or RH between different SOA phase states. A direct way to measure the glass transition temperature, T_g, of SOA substances is to use differential scanning calorimetry (DSC, Zobrist et al., 2008). However, this method requires the removal of the semivolatile aerosol particles from the surrounding gas, which can change the state, shape, or composition of the SOA particles. In contrast, indirect methods allow the sampling of the SOA particles with their surrounding gas. As indirect methods, they do not directly measure T_g, but rather probe the change in the SOA particle mechanical or aerodynamical properties with temperature or RH. These methods include the bounce method that exploits the fact that solid or semisolid particles rebound in an impactor plate, whereas liquid particles adhere. A low pressure impactor (ELPI; Virtanen et al., 2010; Saukko et al., 2012) has been used for this purpose and, recently, the method is extended to work at atmospheric pressures (Bateman et al., 2014). Moreover, the so-called *poke-flow* method (Renbaum-Wolff et al., 2013) and the shatter technique (Schill et al., 2014) use mechanical forces to disturb single particles and the response to the mechanical stress can be linked with a phase state. Finally, the particle phase state can be determined with measurement of their aerodynamic shape factor (Zhang et al., 2015).

Although these methods provide useful insights into the nature of the SOA particles, the SOA aerosol has been sampled and passed through an instrument, in which their environment is altered. This may result in a change in the material properties. In an effort to minimise biases due to these perturbations, I have employed a new optical method that provides a noninvasive probe of shape or structural anisotropy that may be present in solid or highly viscous semisolid SOA particles (Virtanen et al., 2010; Adler et al., 2013; Pajunoja et al., 2014; Schill et al., 2014; Zhang et al., 2015), but that is quickly erased in low viscosity droplets as they relax toward the most energetically favourable state, i.e., spherical droplets that are homogeneous throughout their volumes.

4.2.1. Experimental procedure

The experiments presented here were conducted at the CERN CLOUD chamber (Kirkby et al., 2011; Duplissy et al., 2015) during the CLOUD8 (November–December 2013) and CLOUD9 campaigns (September–November 2014). Each experiment commenced with clean, aerosol-free air (background concentration well below 1 cm^{-3} and low RH (ranging from 5-15%). The SOA particles were generated and grown in the chamber by continuous oxidation of α-pinene (maximum concentration $\sim$600 ppbv) with ozone (maximum concentration $\sim$700 ppbv) to form low volatility oxidised organic compounds. The ozone was monitored with a trace gas monitor (O_3 analyser, Thermo Environmental Instruments, Inc., Model 49C); α-pinene was measured by a PTR-TOF (Proton Transfer Reaction Time of Flight Mass Spectrometer, Ionicon Analytic). SOA nucleation was initiated by injecting α-pinene (10 ml/min) and ozone (1000 ml/min) into the chamber simultaneously for 1 to 7 minutes, depending on the chamber temperature, and then turning on the UV light (Philips TUV 130 W XPT lamp). After 1 minute the α-pinene flow was turned off until the freshly nucleated particles had grown to a mean diameter of approximately 100 nm. This produced a near-monodisperse particle population measured by the Ultra-High Sensitivity Aerosol Spectrometer (UHSAS, DMT). At this point the α-pinene flow was turned back on and continuously injected into the chamber in order to grow the existing particles without inducing further new particle formation. The initial particle burst typically comprised around 50 000–100 000 cm^{-3} α-pinene SOA particles, which rapidly grew to about 20 nm diameter. After the particles had grown to diameters of approximately 600 nm, losses due to dilution, coagulation and the chamber walls had reduced the particle number concentration to 5 000–10 000 cm^{-3}. At this point, the chamber RH was increased to trigger phase transition to liquid phase. These experiments were performed at four different temperatures: -10°C, -20°C, -30°C and -38°C.

In addition to the experiments at low RH, I performed one reference run with a higher starting RH (60–70 %) at −20°C. The nucleation and growth of the α-pinene SOA particles was conducted in the same way as the other experiments but the high RH was kept constant throughout. The purpose of this reference run was to produce a similar size distribution of α-pinene SOA particles as in the viscosity-transition runs, while ensuring that the state of the particles did not change, i.e., remained liquid, so they could not induce depolarisation.

I used this run to verify the depolarisation method for a SOA system and to exclude the possibility of artefacts.

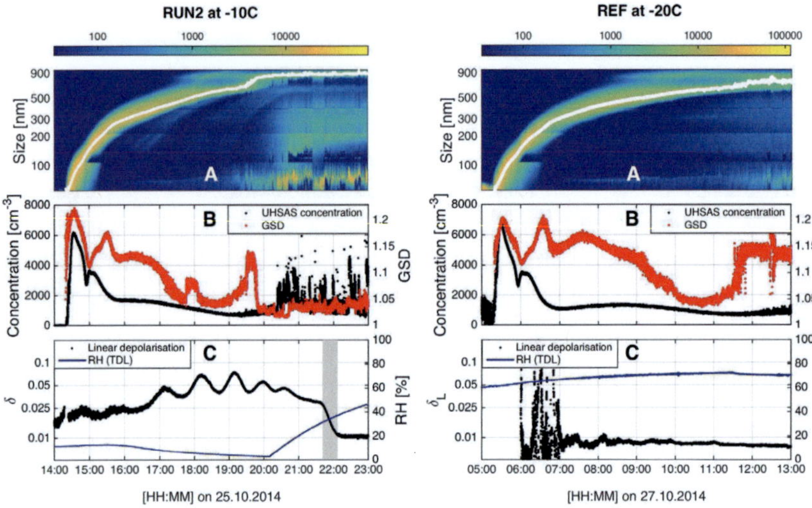

(a) Experiment run at -10°C.　　**(b)** Reference run at -20°C.

Figure 4.7.: Two SOA runs. The particle size distribution and the GMD from a log-normal fit are shown in panel A. Panel B shows the total concentration measured with the UHSAS and the GSD determined from a log-normal fit. Panel C shows the linear depolarisation ratio and RH.

4.2.2. Depolarisation properties of the α-pinene SOA particles

Eight SOA nucleation and growth experiments were performed in the CLOUD chamber using identical initial conditions as described in Sec. 4.2.1. Each experiment resulted in a nucleation event that was followed by rapid growth of the α-pinene SOA particles, producing a narrow particle size mode. Two examples of the growth of a single narrow mode are shown in Figs. 4.7a and 4.7b. I fitted a log-normal distribution to the particle size distribution data to retrieve the geometric mean diameter (GMD) and geometric standard deviation (GSD). The GMD are shown in the first panels, and the GSD together with number concentration in the second panels. From the log-normal fit results, we see that the two runs presented in Figs. 4.7a and 4.7b are almost identical with respect to the particle size distribution, even though they were conducted at different temperature and humidity conditions. Both size distributions are narrow with GSD variations between 1.05–1.2, and show similar total number concentrations. In each case, particles grew beyond 700 nm.

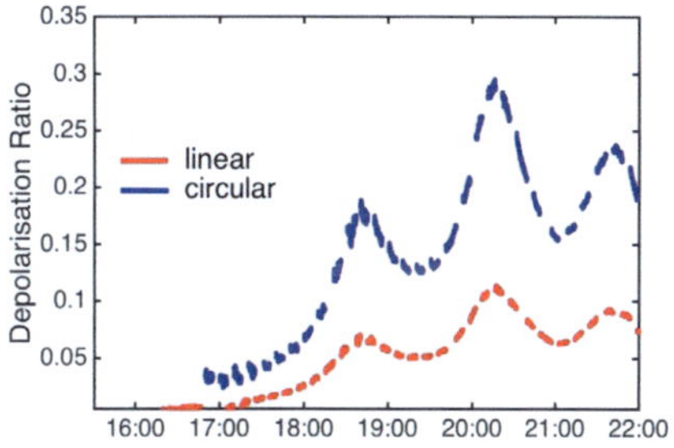

(a) An example of alternating measurements of the linear and the circular depolarisation ratios. Both depolarisation ratios show oscillations with consistent minima and maxima.

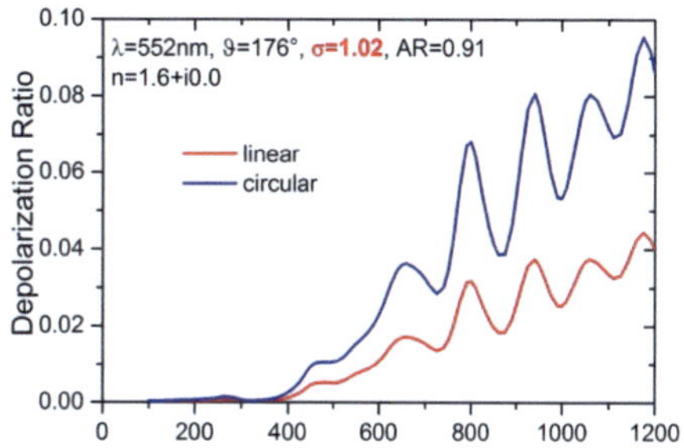

(b) Modelled linear and circular depolarisation ratio using spheroidal model.

Figure 4.8.: Oscillations in the depolarisation signal.

During all the low RH experimental runs a depolarisation ratio above LOD was measured, whereas during the reference run, the depolarisation signal stayed below the depolarisation threshold of 0.01 throughout the growth of the SOA particles. Based on this comparison, we can conclude that at low RH, the measured depolarisation ratio was due to the physical properties of the α-pinene SOA particles and not a result of measurement artefacts. The measured linear depolarisation ratio averaged between 0.01 and 0.1 (Fig. 4.7a), and the circular depolarisation ratio varied between 0.05 and 0.3 (Fig. 4.8a). The circular depolarisation ratio was always higher than linear depolarisation ratio. This is a common feature in depolarisation measurements, and has been discussed in previous work by Mishchenko and Hovenier (1995b) as well as shown in Sec. 4.1.4. In theory, the circular depolarisation ratio is more sensitive to small changes; this sensitivity should provide advantages over linear depolarisation. However, in our case the LOD for circular depolarisation ratio is a factor two higher than that for linear depolarisation ratio. Thus, for this instrument, circular depolarisation offered no clear advantage.

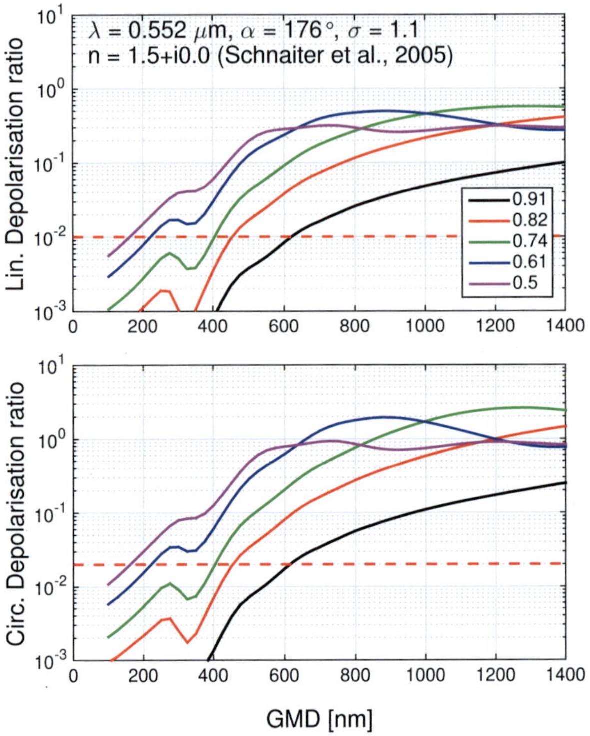

Figure 4.9.: T-matrix calculations for the depolarisation properties of spheroids with a constant GSD of 1.1 and a SOA-like refractive index. The detection thresholds of the SIMONE-Junior instrument (1% for linear depolarisation and 2% for circular) are indicated as dashed red horizontal lines.

A depolarisation ratio that is above the LOD in these experiments suggests the presence of aspherical particles. In this case the magnitude of the depolarisation signal depends on the size and shape of the particles. I modelled the depolarisation properties of SOA particles assuming a spheroidal shape (Fig. 4.9). The increase in the depolarisation at the beginning of the runs can be linked with the growth in the particle size, as can be expected from the T-matrix model. A median particle diameter of 100 nm was sufficient to produce a depolarisation ratio above the detection threshold; the highest depolarisation ratio was measured for 600 nm particles (Fig. 4.7a). The maximum linear depolarisation ratio varied between 0.05 and 0.1, which corresponds to an aspect ratio of 0.8 to 0.74 according to the T-matrix model for spheroidal particles. The aspherical shape of the SOA particles is likely a result of coagulation that took place at the early stages of the experiment, when the particle number concentrations were high. Later in the experiments, coagulation became insignificant and the condensational growth drove the coagulated particles to become increasingly spherical, which is also indicated by the measurements, as a slight decrease in the depolarisation ratio as the particle median diameter grew beyond 600 nm.

In all the experiments, both linear and circular depolarisation ratio showed oscillation (Fig. 4.8a). Oscillation in the scattering signals is a typical feature of spherical particles with a narrow size distribution and can be well explained with the Mie theory. Therefore, it is surprising to observe oscillation in the depolarisation signal that should be the measure of particle non-sphercity and to my knowledge such a behaviour have not been previously reported. A comparison with a spheroidal model, showed that the oscillation behaviour is a feature of the narrow size distribution (in the model a GSD between 1.02 and 1.05), and can be produced even when the particles are not spherical (model particles have an aspect ratio of 0.9).

4.2.3. Observation of shape transition

The depolarisation signal in the experiments showed that the viscous α-pinene SOA particles were non-spherical when the nucleation and growth of the particles occurred at low RH. As the RH increases, the highly viscous SOA particles start to take up water. The water uptake is slow, and proceeds gradually with increasing RH (Mikhailov et al., 2009; Zobrist et al., 2011). This is seen by the relatively slow shape transitions observed as the RH increases (Fig. 4.7a). In the aqueous phase, the surface area is minimised to reach an energetically favourable state. The minimisation of the surface area results in a shape change from aspherical to spherical. This change was observed in the depolarisation ratio at the end of the low-RH runs, as the depolarisation ratio decreased towards the instrument LOD. I determined the transition RH range from the measurements; the start of the transition was detected when the depolarisation ratio started to decrease significantly. The end of the transition was detected when the depolarisation ratio reached a constant level below the detection threshold (highlighted with grey in Fig. 4.7a). The RH value at the time the depolarisation had decreased to a constant, zero level, describes the transition to an optically spherical shape. I label this RH value the viscosity/shape-transition RH.

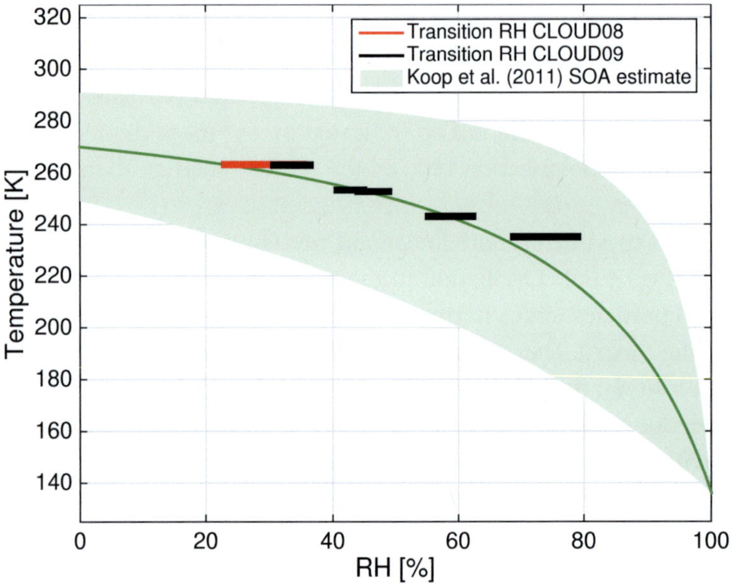

Figure 4.10.: Transition RH at different temperatures. The horizontal lines show the RH range for the transition and the the width of the lines represent the temperature uncertainty of 2 K. The green curve is the generic SOA estimate from Koop et al. (2011) and the shaded area represents the upper and lower boundary for the estimate.

Figure 4.10 summarises the complete RH range, where the viscosity/shape-transition were observed for sub-zero temperatures. At 10°C a depolarisation signal over LOD was not observed, suggesting that the particles were already in a liquid or low-viscous state at an RH of 12 %. The measured viscosity/shape-transition RH increases with decreasing temperature and has near linear dependency on the temperature. At −10°C the transition occurred at RH around 35 % and at −38°C the transition was observed at around 80 % RH.

It should be highlighted that the presented method does not directly measure the change in the viscosity but provides the critical RH, where the water uptake is significantly enhanced and the particles start to relax rapidly due to the change in their viscosity. During this process the viscosity changes of the α-pinene SOA particles are not limited by diffusion on water into the particles as the assumed water diffusion timescales can be expected to be in the order of seconds (Price et al., 2015; Lienhard et al., 2015). We can now use the observed relaxation time (of the order of 10 minutes) to provide a rough estimate of the viscosity of the α-pinene SOA particles during their viscosity/shape-transition. However, it has to be kept in mind that the particle viscosity and the relaxation timescale are affected by the constant increase in the chamber RH by about 0.5% per minute, which increases the uncertainty of our estimate.

For the estimation of the particle viscosity, I used the method described in Pajunoja et al. (2014). I assumed a primary particle size of 400 nm in an SOA aggregate and a surface tension of 35 mN/m (Pajunoja et al., 2014). With these numbers I get an estimated viscosity of 10^7 Pa s. This value describes the viscosity during the viscosity/shape-transition period. Previous studies (e.g. Renbaum-Wolff et al., 2013; Bateman et al., 2014; Kidd et al., 2014; Pajunoja et al., 2014) have studied the viscosity of α-pinene SOA particles at room temperatures. The studies have found that α-pinene SOA particles are generally in liquid state already at RH$>70\%$ (Song et al., 2015). At colder temperatures we can expect the transition from viscous to liquid state is shifted to higher RH (Wang et al., 2014). As the measured viscosity/shape-transitions generally took place at RH$<70\%$ the comparison with literature would suggest that albeit the SOA particles were spherical, they probably were not in liquid state after the transition. Therefore, the measured viscosity/shape-transition would describe the change in the viscosity within a semi-solid material.

I compared the obtained results with the Koop et al. (2011) generic SOA estimate that predicts the glass transition temperature as a function of RH for a broad variety of SOA (Fig. 4.10). Overall the results agree well with the generic SOA estimate, and significantly reduce the uncertainty in that estimate. Moreover, the results suggest the relationship between the transition RH and the temperature is more linear than predicted by Koop et al. (2011). This may be especially relevant at high RH, where we can expect that the transition might take place at a much higher temperature (around 220 K) than previously predicted.

4.3. Summary and conclusions

In the beginning of this chapter we discussed atmospheric lidar measurements and stated that a requirement for successful lidar retrievals is to understand the light scattering properties of the particles of interest (the forward problem). To help answer this problem, I investigated the backscattering depolarisation properties of two aerosol systems: dust particles that are naturally complex and α-pinene SOA particles, whose complexity was induced via viscous phase state. In the case of dust particles, I studied the size-dependence of the depolarisation ratio with well-defined and mono-modal dust size distributions. I found out that the depolarisation properties of the different dust samples were similar and only dependent on the particle size. The strongest size-dependence was observed for fine-mode particles as their depolarisation ratios increased almost linearly with particle median diameter from 0.03 to 0.3, whereas the coarse-mode particle depolarisation values stayed rather constant with a mean linear depolarisation ratio of 0.27. This clear separation in the fine- and coarse-mode depolarisation properties can be used in lidar retrievals to improve algorithms to separate the two dust modes.

Also, it seems that generally discriminating dust from other aerosol particles might be more straightforward due to their stable and relatively high depolarisation ratio. We learned in chapter 2 that the typical δ_L values measured by lidar methods for atmospheric dust in field campaigns are generally around 0.3, independent of the source region of the dust, and show little variation. Similarly, the measurement results presented in this thesis show a robust depolarisation behaviour that had little variation between the different samples. Therefore, it can be concluded that the depolarisation properties of dust particles might be universal and largely independent of the origin of the dust.

If dust particle depolarisation properties indeed seem universal, it could be a result of the particle complexity. Several observations support this statement: (i) Complexity of dust particles (as seen in Fig. 4.1) certainly is an universal feature and mineralogical studies have shown that the composition of natural dust particles is inhomogeneous (Nousiainen, 2009), which further contributes to complexity. We learned in Chapter 2 and in Sec. 3.2.3 that particle complexity removes shape specific scattering features. Similarly, (ii) increasing particle complexity in scattering models seems to always result in similar depolarisation behaviour (Nousiainen, 2015; personal communication), which (iii) was also observed in the measurements. Note, that the complex dust particles also fulfilled the Hovenier and Mishchenko theorem, even if dust particles cannot assumed to be mirror symmetric.

In the case of SOA particles, by adding a small degree of complexity, here a small deformation from the spherical shape, altered the depolarisation properties of those particles so that the phase state was detectable by depolarisation measurements. These experiments demonstrated, how sensitive the backscattering depolarisation ratio is for deviations from the isotropic spherical shape and showed the potential of light scattering measurements to unperturbedly probe small changes in the particle shape. However, both the experiments with the complex dust particles and with the viscous SOA demonstrated the general feature of the depolarisation measurements: the depolarisation ratio approaches zero at small sizes, which makes detecting small particles challenging. In the experiments presented here, both the SOA and the dust particles had to be $> 100\,\text{nm}$, before their depolarisation signals were measurable.

5. Laboratory investigations on the angular light scattering of complex ice particles

In this chapter we move from aerosol angular light scattering properties to investigate light scattering by ice particles. Unlike accumulation and coarse-mode aerosol particles that have been transported from Earth's surface to the troposphere, ice particles are formed in the atmosphere through complicated processes, like homogeneous or heterogeneous ice nucleation or in ice multiplication processes. These processes take place over a wide range of environmental conditions, further influencing the ice crystal growth. Therefore, the key question is, how are ice particle morphology and complexity linked to these microphysical processes and environmental conditions.

We learned that ice particle small-scale complexity is an important feature in natural ice crystals. The observations of ice crystal complexity have motivated an increasing number of laboratory studies investigating ice crystals under electron microscopes (Neshyba et al., 2013; Magee et al., 2014; Ulanowski et al., 2014). According to these studies, ice particle small-scale complexity is a common feature and there are indications that it is linked to sublimation. However, results have to be taken with care because of possible artefacts due to the substrate, the low pressure in the microscope and the electron beam.

There is a clear need for further dedicated studies to reveal the origin of ice particle small-scale complexity. In-situ investigations in natural cirrus have tried to resolve this, but are hampered due to complicated and uncontrolled atmospheric processes (Ulanowski et al., 2014). Therefore, controlled laboratory studies could provide answers to this question, as the ice particle history can be tracked down to the ice formation. In this chapter, we will investigate the origin of ice crystal complexity for the case of hexagonal columns and near-spherical ice particles. The hexagonal columns represent the typical ice crystal shape in in-situ formed cirrus at temperatures below -40°. The near-spherical ice particles represent ice particles in convective clouds that have a liquid origin and are formed at temperatures above -40°. We will also investigate the angular light scattering properties of complex ice particles, their near-backscattering depolarisation properties and the effect of ice particle complexity to discriminating ice particle from water droplets.

5.1. The origin of ice crystal complexity in cirrus

The origin of ice crystal complexity in the case of hexagonal columns was studied in a series of AIDA chamber studies called the **R**ought **ICE** campaigns between 2012-2014 (RICE01-RICE03). The experiments were led by Dr. M. Schnaiter and the results were published in Schnaiter et al. (2016). In my PhD project, I was involved in the last two RICE-campaigns, and therefore a brief summary of the main results is given in Sec. 5.1.1. During these campaigns I investigated the light scattering properties of complex and pristine ice crystals by measuring their angular scattering functions (Sec. 5.3) and their near-backscattering depolarisation properties (Sec. 5.4). I also extended the complexity studies with a series of chamber simulations for convective systems (Sec. 5.1.2). These studies were performed at the AIDA chamber during the RICE03 campaign (Järvinen et al., 2016c) and at the CERN CLOUD chamber during the CLOUD9 campaign (unpublished results).

5.1.1. Ice crystal complexity in case of hexagonal ice particles

The ice crystal complexity of different heterogeneously and homogeneously nucleated ice particles were studied in chamber experiments at temperatures between -40° and -60°. A novel experimental procedure was developed to grow and sublimate ice crystals under specific super- and sub-saturated ice conditions. The initial cloud formation took place by heterogeneous nucleation on soot or dust aerosol or by homogeneous nucleation in SA solution droplets. As the initial growth of the ice particles depends on the specific ice nucleation properties of the aerosol used, the growth conditions in terms of supersaturation cannot be controlled. Therefore, a subsequent sublimation period was applied to remove the ice particle surface characteristics from the initial growth. After processing the ice particles with sublimation, a re-growth was initiated and maintained at a set ice supersaturation levels. A more detailed description of the experimental procedure can be found in Schnaiter et al. (2016). Altogether 45 of sublimation and re-growth cycles were made to study the small-scale structure of the ice particles after growth in distinct ice supersaturated conditions with ice saturation ratios S_{ice} from close to 1 to 1.4.

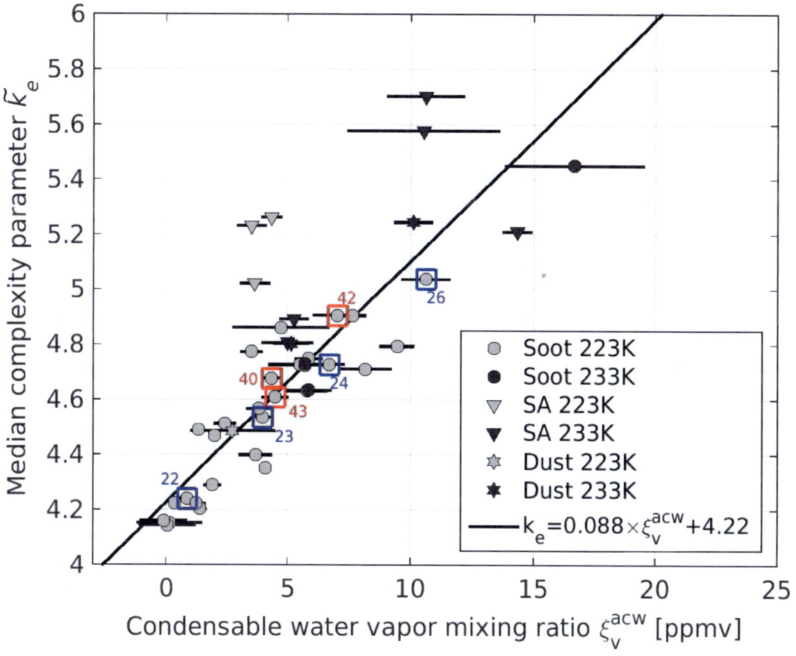

Figure 5.1.: The median ice crystal small-scale complexity parameter $\bar{k}_e$ deduced from the SID-3 patterns plotted against the available condensable water vapour mixing ratio, ζ_v^{acw}. From Schnaiter et al. (2015).

The results from these experiments are shown in Fig. 5.1. The ice crystal small-scale complexity was determined from the SID-3 scattering patterns using the procedure described in Sec. 3.2.3. It was found that the driving factor for the crystal complexity is the ice crystal growth rate that is determined through S_{ice} and temperature. These variables determine the amount of water molecules that are free to condense to the ice phase (Fig. 5.2), the so-called available condensable water mass. Therefore, the ice saturation ratio S_{ice} that was used to grow the ice particles in each re-growth experiment was transferred into the available condensable water vapour mixing ratio, ζ_v^{acw} that is given on the x-axis in Fig. 5.1. Here, ζ_v^{acw} was calculated as the difference between the measured interstitial water vapour mixing ratio ζ_v and the saturated mixing ratio ζ_v^s for the given pressure and temperature conditions inside AIDA. The saturated mixing ratio, ζ_v^s, was calculated using the parametrisation of the ice vapour pressure by Murphy and Koop (2005). Now, ζ_v^{acw} represents the mixing ratio of excess water molecules that can deposit to the solid ice face.

From Fig. 5.1 we see that the small-scale complexity is directly proportional to the ζ_v^{acw}. A linear regression model can be fitted to the data to get a functional dependency between the $\bar{k}_e$ and ζ_v^{acw}:

$$\bar{k}_e = 0.088 \times \zeta_v^{acw} + 4.22, \tag{5.1}$$

where ζ_v^{acw} is given in ppmv. Furthermore, we see that the correlation shown in Eg. 5.1 does not in general depend on the aerosol type or on the nucleation mode. Few exceptions are seen in the case of homogeneous nucleation, and these exceptions show higher $\bar{k}_e$ value compared to the corresponding experiments with heterogeneous initial ice formation. A possible explanation for this is a formation of concentrated H_2SO_4/H_2O residuals on the ice particle surface, which affects the ice crystal growth (Schnaiter et al., 2016).

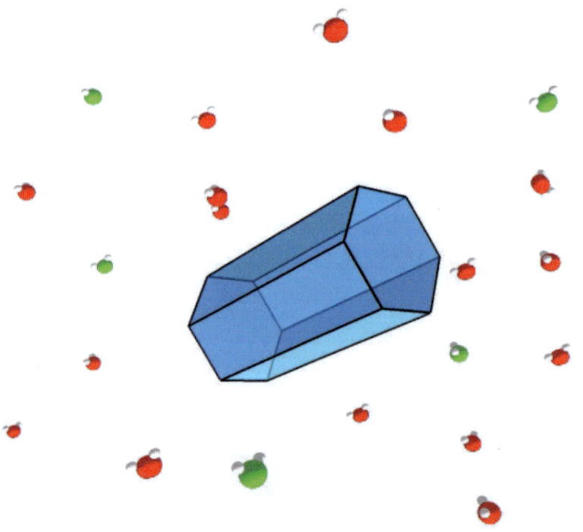

Figure 5.2.: Illustration of the definition of ζ_v^{acw}. The "red" water molecules are free to deposit to the solid phase. If the number of these molecules is too high, they cannot find a place at the prismatic edges and can contribute to small-scale complexity.

5.1.2. Ice crystal complexity in case of near-spherical ice particles

Convective systems are an important source of ice particles in the upper troposphere (e.g. Jensen et al., 1996; Gayet et al., 2012a; Frey et al., 2011) and the lower most stratosphere (Reus et al., 2009). Ice particles found in the anvil outflows are usually formed in the lower and warmer part of the convective cell, and therefore, their microphysical and optical properties differ from in-situ formed ice particles (e.g. McFarquhar and Heymsfield, 1996; Lawson et al., 2003; Connolly et al., 2005; Lawson et al., 2010; Frey et al., 2011). In mid-latitude convective systems, supercooled liquid water droplets have been observed to survive down to a temperature around -37°C (Rosenfeld and Woodley, 2000), where homogeneous freezing of the supercooled droplets occurs. In vigorous convective systems, homogeneous freezing happens in a narrow time interval producing a large amount of small ice crystals, so called frozen droplets (Heymsfield and Sabin, 1989; Philips et al., 2007; Lawson et al., 2010). The ice particle concentrations at the top of convective systems can reach several tens per cubic centimetre, with the effective diameter of the ice particles staying below 50 µm (Lawson et al., 2010; Gayet et al., 2012a; Stith et al., 2014).

Frozen droplets are spherical or quasi-spheroidal small particles that can be identified from measurements with cloud particle imagers (CPI), which were used in studies of Lawson et al. (2003), Gayet et al. (2012a) and Stith et al. (2014). The optical resolution of CPIs typically range from about 2 to 5 µm depending on the probe model and aircraft speed, i.e. too coarse to resolve the fine structure of small ice particles. What is detected as apparently spherical ice particle, may actually be a complex polycrystal or a severely roughened ice particle (Ulanowski et al., 2004). Therefore, measuring spatially resolved light scattering patterns (SID-3 method, Sec. 3.2) instead of the particle image can bring new insight to the microphysical nature of the frozen droplets. In the case of frozen droplets, the SID-3 images offer two types of information: (i) the sphericity of the ice particles and (ii) the small-scale complexity or surface roughness.

I simulated the droplet freezing process in the cloud simulation chambers AIDA and CERN CLOUD, where frozen droplets were grown and sublimated under controlled conditions. During the growth and sublimation cycles, the size, shape, complexity and angular light scattering properties of the frozen droplets were investigated to understand the link between environmental, microphysical and optical properties. The experiments were started with a clean chamber at -30°C. As seed aerosol I used SA solution droplets. At the AIDA chamber I simulated updrafts with moderate updraft speeds of 2 m s^{-1} to 7 m s^{-1}. The experiments were complimented with a data set from the CERN CLOUD chamber, where high updraft speeds of ~30 m s^{-1} were simulated.

Fig. 5.3 shows a droplet freezing experiment conducted with an initial number concentration of 12 cm^{-3} SA solution droplets. The expansion was started at experiment time 0 s, as indicated by the start of the pressure decrease in Fig. 5.3, panel a. The cooling rate at the beginning of the expansion was −2.5 K min^{-1}, but in the course of the expansion, the heat flux from the chamber walls reduced the cooling rate. At experiment time 83 s, water saturation was reached (dashed blue line in panel b), and a cloud of supercooled liquid droplets was formed, indicated by the rapid increase in the forward scattering intensity (black line in panel c). Moreover, a zero depolarisation ratio was measured indicating the presence of spherical particles in this period (red line in panel c). The cloud particles were detected by the PPD-2K instrument after experiment time 100 s, when they have grown to diameters above 7 µm (panels d and e). In this first phase of the experiment, the 2D diffraction patterns of supercooled droplets recorded by the PPD-2K showed concentric ring pattern (Fig. 5.4 i) with v_{az} below the threshold value of 1×10^{-5} (Fig. 5.3 panel e).

The cooling was continued until the homogeneous freezing threshold around 237 K was reached at the experiment time 132 s. This led to a rapid glaciation of the cloud through homogeneous freezing of the supercooled droplets. Just before freezing, the liquid droplets had reached a median diameter of 14 µm. The glaciation of the cloud led to an increase of the ice water content (IWC), as indicated by the difference between the total water (MBW, black line in panel b of Fig. 5.3) and the interstitial water (TDL, solid blue line in panel b of Fig. 5.3). At the same time, the depolarisation ratio (panel c) started to depart from zero, and reached a maximum of ~ 0.3 at 200 s after the droplets were fully depleted. The optical size of the frozen droplets did not significantly differ from the droplet size of the initial liquid and, therefore, the glaciation is not visible in the PPD-2K size distribution in panel d. Yet, the variance analysis (quantification of particle sphericity, see Sec. 3.2.2 for details) clearly showed an increase in the v_{az} during the mixed-phase conditions. After the full glaciation around experiment time 190 s, a sharp transition to v_{az} above the threshold value of 1×10^{-5} was observed, indicating the presence of non-spherical particles.

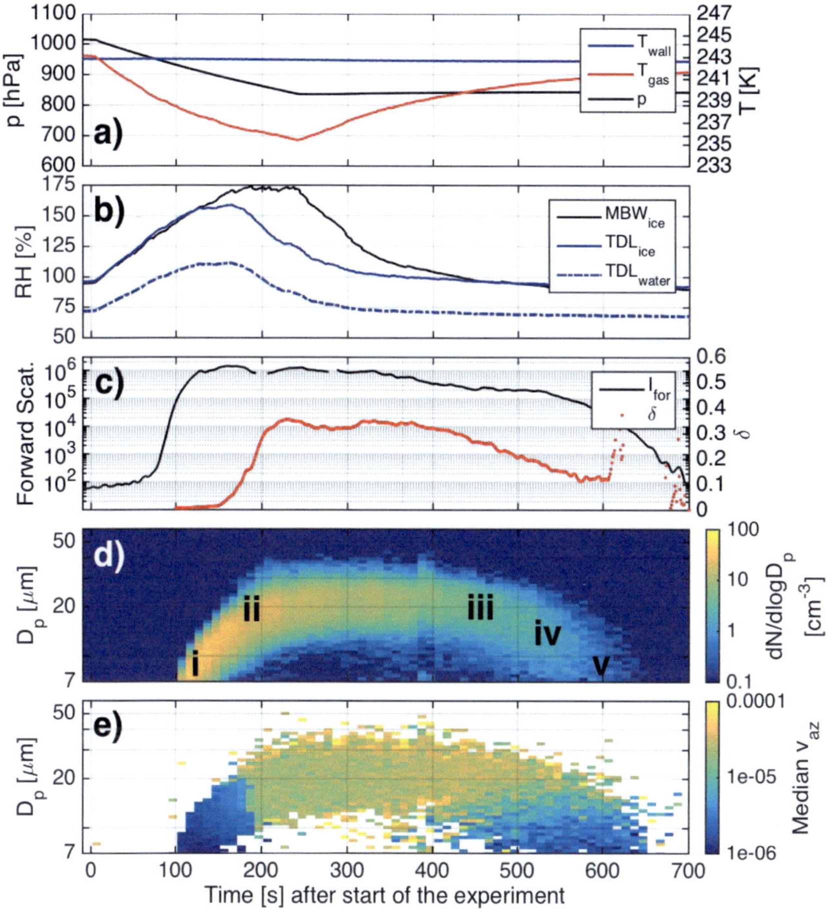

Figure 5.3.: A droplet freezing experiment initiated with 12 cm^3 sulphuric acid aerosol and an initial cooling rate of $-2.5\,K\,min^{-1}$. Panel a) shows the pressure of the chamber (black line) as well as the chamber and wall temperatures (red and blue lines, respectively). Panel b) shows the total water measured with MBW (black line) and the interstitial water with respect to ice (blue solid line) and water (blue dashed line) measured with TDL. The forward scattering intensity (black line) and the depolarisation ratio was measured for cloud particles in the middle of the chamber and is shown in panel c). Panel d) shows the PPD-2K size distribution and panel e) the size-segregated median variance of the 2D scattering patterns. The expansion of the chamber volume was started at experiment time 0 s.

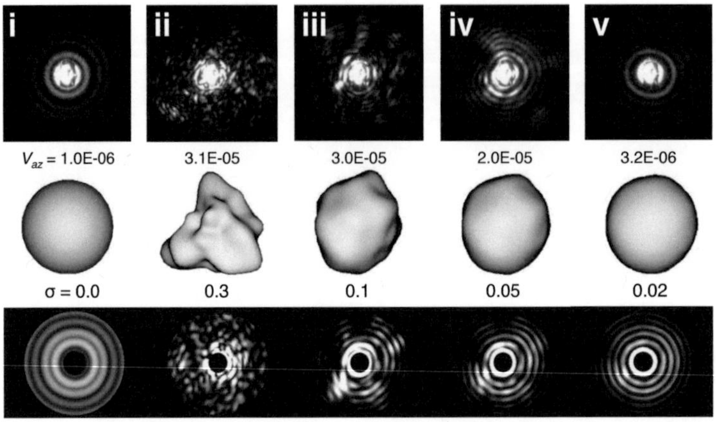

Figure 5.4.: Simulation of the deformation of the surface of near-spherical ice particles with the Gaussian random sphere model. The surface was altered with a relative deformation between 0 and 0.3 (middle row). The resulting 2D scattering patterns are shown in the third row. Particles i-v (first row) were measured with the PPD-2K during the experiment shown in Fig. 5.3.

The PPD-2K scattering patterns of ice particles during the growth in mixed-phase conditions and later through vapour deposition were dominated by speckles (Fig. 5.4 ii), indicating a significant degree of complexity. I measured a complexity parameter, k_e, of 6.5 for frozen droplets formed from liquid phase. These values were significantly larger than what was measured for vapour grown ice crystal at cirrus temperatures (Sec. 5.1.1). This could indicate that the small-scale complexity of liquid-origin ice particles could be severely enhanced compared to in-situ grown ice crystals. Larger-scale complexity, e.g. riming, is frequently found in mixed-phase cloud (e.g. Ono, 1969), but due to instrument limitations, small-scale complexity have not been studied. Therefore, field measurements in mixed-phase regions with SID-type instruments would be needed to validate this assumption.

The growth of the frozen droplets was stopped after a median diameter of 22 µm was reached. In the third phase of the experiment, the frozen droplets were forced to sublimate under ice sub-saturated conditions. The sublimation was seen in the PPD-2K diffraction patterns, as ring-like patterns started to emerge, and these patterns became more concentric towards the end of the sublimation. The emergence of the rings can be linked with the decrease in the crystal complexity, as sharp edges on the ice particle have a higher saturation vapour pressure and, therefore, tend to sublimate first.

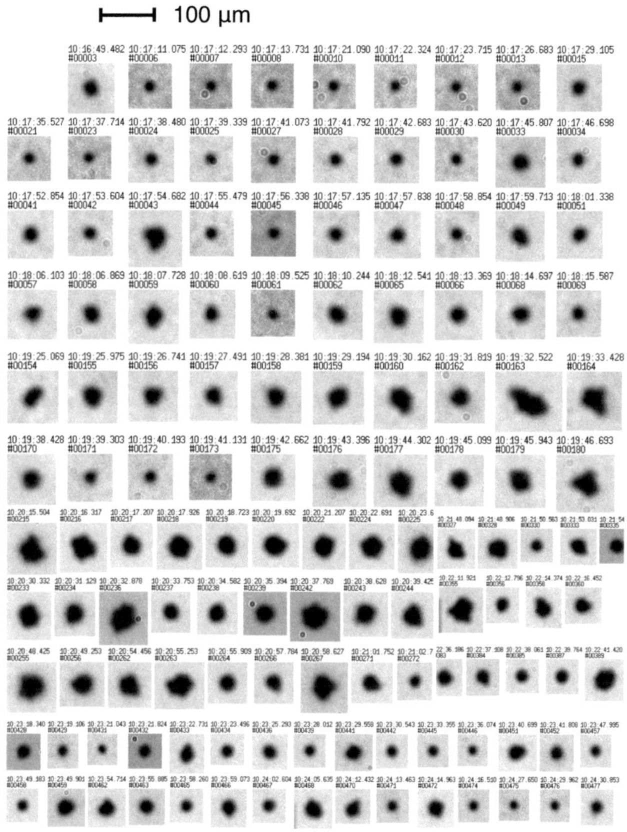

Figure 5.5.: Examples of PHIPS-HALO images of frozen droplets in chronological order.

This is also seen in the v_{az} (Fig. 5.3 panel e); the v_{az} slowly decreased to values below the threshold value, and at the end of the sublimation period, the v_{az} of the sublimating frozen droplets was almost equivalent to that of liquid droplets (compare the v_{az} of the liquid droplet (i) and sublimating frozen droplet (v) in Fig. 5.4). However, these spherical particles cannot be liquid droplets, as the temperature during sublimation period of the fully glaciated cloud stayed well below -30°C. Because the frozen droplets at the end of the sublimation period cannot be distinguished from spherical droplets according to the variance analysis, I call these ice particles *optically spherical*. The angular scattering measurements and the depolarisation measurements later confirm this statement.

The difference between optically spherical and near-spherical ice particles is well depicted in Fig. 5.5. It shows the PHIPS-HALO images of frozen droplets during the experiment in a chronological order. At the beginning the ice particles seem almost perfectly spherical, although based on the PPD-2K variance analysis and the SID-3 complexity analysis we know that these particles were highly distorted. The distortion lies in the microstructures of these particles and, therefore, cannot be seen from the PHIPS-HALO images with restricted resolution. Only after certain growth, the non-spherical nature of these particles is exposed. In the sublimation the ice particles rather quickly loose the clear aspherical features and become again near-spherical. Now, both the variance analysis and the complexity analysis agree that the near-spherical particles are also optically spherical. Therefore, although the first and the last PHIPS-HALO image in Fig. 5.5 look almost identical, their light scattering properties are very different, which highlights the need of sophisticated measurement techniques in investigations of the microphysical nature of small ice particles.

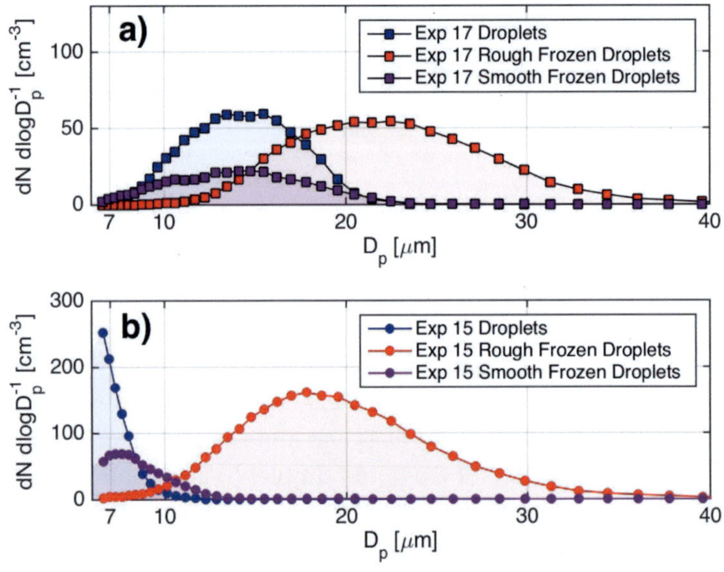

Figure 5.6.: Averaged size distributions of supercooled droplets before freezing, rough frozen droplets at the end of the growth period, and of optically spherical frozen droplets. Experiment (a) was started with an aerosol concentration of 12 cm^3 and experiment (b) with 989 cm^3.

At the end of the sublimation period the particle size distribution of the optically spherical frozen droplets was found to be close to that of the supercooled droplets at the beginning of the experiment (Fig. 5.3). To investigate, whether the size of the optically spherical frozen droplets corresponds to that of the liquid droplets before freezing, I performed experiments with varying initial mean droplet sizes. The droplet size can be controlled by changing the initial aerosol concentration. Fig. 5.6 shows size distributions of liquid droplets, frozen droplets after freezing and initial growth and optically spherical frozen droplets in two different experiments: one with a low initial aerosol concentration of $12\,\mathrm{cm}^{-3}$ and the other with a higher initial aerosol concentration of $989\,\mathrm{cm}^{-3}$. We see that in both cases the size of the optically spherical frozen droplets is comparable to the initial liquid droplet size. Therefore, it can be concluded that the liquid droplets keep their spherical form in the freezing process, but the spherical shape is quickly distorted under the rapid depositional growth at near-water saturated conditions.

Simulation of droplet freezing in a thunderstorm

The thickness of the frost layer or the degree of the deformation of the frozen droplet depends on the growth conditions, e.g. the duration of the mixed-phase cloud. How long the mixed-phase cloud can persist, is largely governed by the updraft speed. In thunderstorms, very high updraft speeds exceeding $15\,\mathrm{m\,s}^{-1}$ have been reported (e.g. Marshall et al., 1995; Deierling and Petersen, 2008), which can lead to high cooling rates that can freeze the supercooled liquid droplets quasi-simultaneously. Under such conditions, the mixed-phase conditions can be regarded as almost non-existent and the frozen droplets cannot efficiently grow through the Bergeron-Findeisen process, which lead to the persistence of the near-spherical shape.

I simulated thunderstorm updrafts of $\sim 30\,\mathrm{m\,s}^{-1}$ at the CERN CLOUD chamber. Figs. 5.7a and 5.7b show particle size distributions and the results from the PPD-2K variance analysis for simulation experiments with $2000\,\mathrm{cm}^3$ and $100\,\mathrm{cm}^3$ initial SA aerosol concentrations, respectively. In both cases the mixed-phase lasted for only a few seconds and is hardly recognisable in the evolution of the size distributions, i.e. the SA aerosol was activated as cloud droplets that froze almost instantaneously. We see a small difference in the size distributions, as the initial aerosol concentration correlates negatively with the ice particle size. Yet, the variance analysis reveals a significant difference in the ice

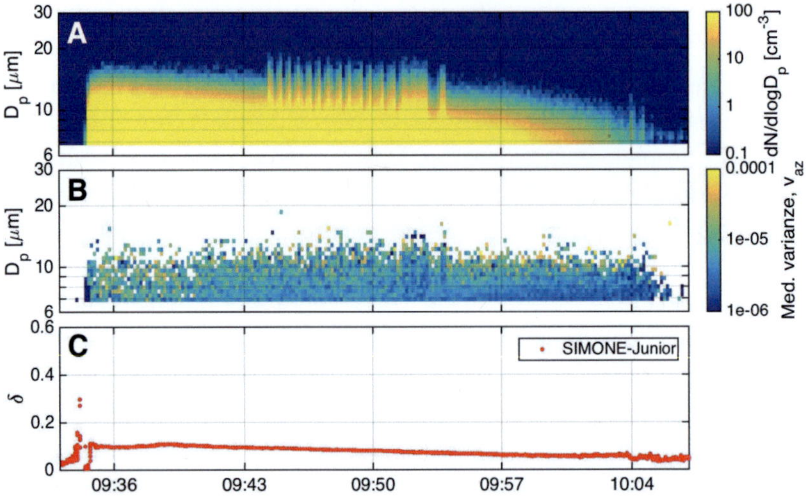

(a) Initial SA aerosol concentration of 2000 cm^3.

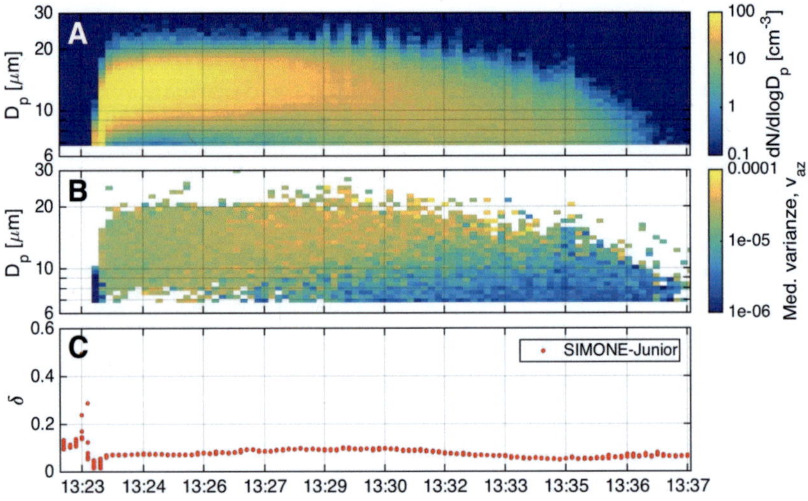

(b) Initial SA aerosol concentration of 100 cm^3.

Figure 5.7.: Simulated thunderstorm updrafts with varying initial aerosol concentrations. Panel A shows the PPD-2K size distribution, panel B the results of the variance analysis, and panel C the linear depolarisation ratio measured by SIMONE-Junior.

particle microphysics: in the case with 2000 cm^3 initial SA aerosol concentration, the variance histogram (Fig. 5.7a, panel B) is dominated by low v_{az} values (blue colours) throughout the experiment, whereas in the case of 100 cm^3 initial concentration we see in the sublimation period a clear transition from non-spherical to spherical (from yellow to blue colours) (c.f. Fig. 5.3).

These two simulation experiments nicely demonstrate the effect of updraft speed and aerosol concentration on the complexity of frozen droplets. With high updraft speeds and high aerosol concentrations (I found a threshold concentration of about 350 cm^3), all droplets freeze simultaneously and create a large ice particle surface, thus preventing the formation of a significant frost layer of the particles. If, on the other hand, the aerosol concentration is low enough, a significant amount of condensable water vapour remains in the system after freezing, which deposits on the ice particle surface and eventually builds a frost layer.

5.2. The influence of ice crystal complexity on the detection of ice particles

Ice particles are usually separated from liquid cloud particles in mixed-phase clouds based on their aspherical shape. Several cloud instruments use optical quantities to determine the fraction of aspherical particles that is then defined as the ice fraction. The v_{az} measured with PPD-2K can be converted into aspherical fraction by applying the v_{az}^{thr}. Furthermore, we learned that aspherical particles have polarisation properties that are distinguishable from spherical particles. This fact is used in airborne cloud particle spectrometers, NIXE-CAPS-DPOL (Meyer, 2012; Luebke et al., 2015) and CAS-DPOL (Baumgardner et al., 2014). These probes measure the cross-polarised component of the backward scattered light for single particles and the ratio of cross-polarised light to the forward scattering light intensity can be converted into aspherical fraction.

Fig. 5.8 shows the aspherical fraction as a function of experiment time in the experiment with near-spherical ice particles (Fig. 5.3). The aspherical fractions were determined from the PPD-2K using the variance analysis in the size range of 8-50 µm and, as a comparison, from NIXE-CAPS-DPOL and from CAS-DPOL using single particle polarisation information in the same size range. Both of the methods show zero aspherical fraction during the liquid phase and a steep increase in the aspherical fraction during the glaciation process. After the full glaciation, the aspherical fractions determined from the PPD-2K and from the CAS-DPOL varied between 0.9 and 1. Only NIXE-CAPS-DPOL shows somewhat lower aspherical fractions around 0.9, which can be traced back to the particle size range <20 µm, where the polarisation signal weakens and thus the ice crystals must have a distinct asphericity to be classified as aspherical. In the sublimation phase of the experiment, the methods show a similar decrease in the aspherical fractions, indicating an increasing presence of smooth sublimating frozen droplets.

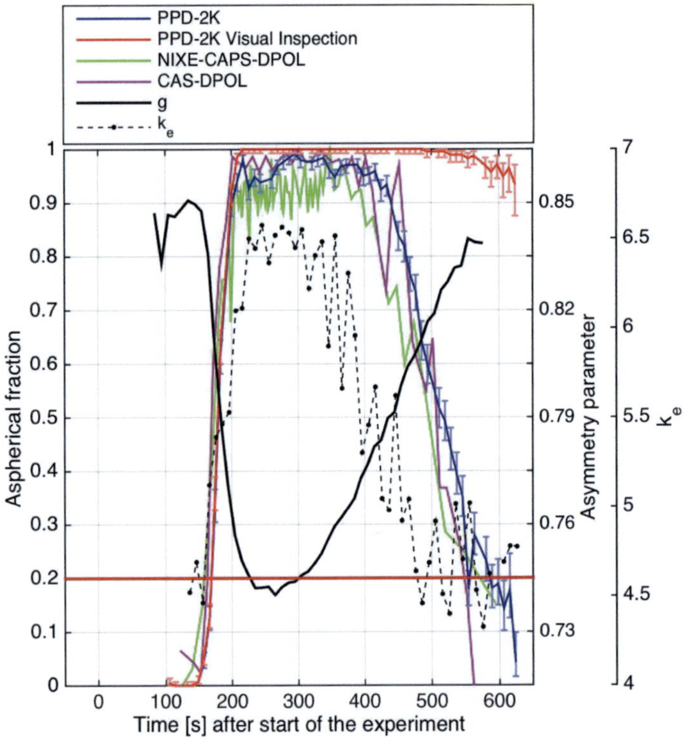

Figure 5.8.: Aspherical fractions during the experiment shown in Fig. 5.3: PPD-2K using an automated routine (blue line) or after applying a manual cross-check (red line), NIXE-CAPS-DPOL (green line) and CAS-POL (magenta line). Also shown is the asymmetry parameter g (black solid line) determined from PN measurements and the complexity parameter k_e (black dashed line) determined from SID-3 measurements.

Fig. 5.8 show that the automated discrimination of aspherical particles clearly correlates with the ice particle complexity in the case of the near-spherical ice particles. At the end of the sublimation period, when the particle complexity was at its lowest, the automated algorithm of PPD-2K and the polarisation based measurements would have misclassified 80 % of the ice particles as droplets. However, the diffraction patterns of PPD-2K still contain information from the ice phase, even if the ice particles seemed to be almost perfect spheres. For example, Fig. 5.4 (v) is a diffraction pattern of a slightly deformed ice sphere. In the PPD-2K this ice particle shows somewhat elongated ring pattern and, therefore, can be identified as an ice particle. Based on its polarisation signal, the same particle would be detected as spherical according to the polarisation method. Thus, I performed a visual inspection of the PPD-2K scattering patterns (red line in Fig. 5.8) and compared these results to the automated routine. As a result of the visual inspection a 100% aspherical fraction was measured during the growth of the ice particles and even at the end of the sublimation period, only 5% of the ice particles were misclassified as droplets, whereas the automated routines would have misclassified 80% of the ice particles as droplets.

The results show that optically spherical ice particles in the atmosphere can lead to misinterpretation of the ice fraction. These particles are especially important in mixed-phase clouds and convective systems, where droplet freezing frequently occurs. The PPD-2K scattering patterns were sensitive to small deviations from the spherical shape and we were able to detect majority of the near-spherical and smooth frozen droplets with the manual inspection, whereas the polarisation based methods led to larger misclassification. However, it has to be kept in mind that the SID-type of instruments have a limited imaging rate of 30 kHz and, therefore, a restricted sampling efficiency compared to the NIXE-CAPS-DPOL and CAS-DPOL.

5.3. The influence of ice crystal complexity on the angular scattering function

I investigated the angular light scattering properties of the roughened hexagonal columns and the frozen droplets with two polar nephelometers: with an established airborne polar nephelometer (PN) from University Blaise Pascal (Gayet et al., 1997; Crépel et al., 1997) and with the PHIPS-HALO instrument.

First we look into the scattering properties of the complex hexagonal columns. Fig. 5.9 shows three angular scattering functions that were measured during experiments highlighted with red in Fig. 5.1. The PHIPS-HALO measurements are marked with squares and the PN measurements with crosses. Although the measured columns in experiments 43, 40 and 42 show increasing complexity from 4.6 to 4.9 (Fig. 5.1), we observe no change in the angular light scattering function: the measured angular scattering function is smooth with a high degree of backscattering. It seems that after a certain complexity threshold is achieved, the angular light scattering is not sensitive to small-scale complexity, whereas the speckle analysis of the spatially resolved forward scattering patterns still showed differences. In the case of hexagonal particles, a threshold complexity parameter of 4.6 was defined to distinguish between pristine columns and complex columns that show uniform angular light scattering behaviour.

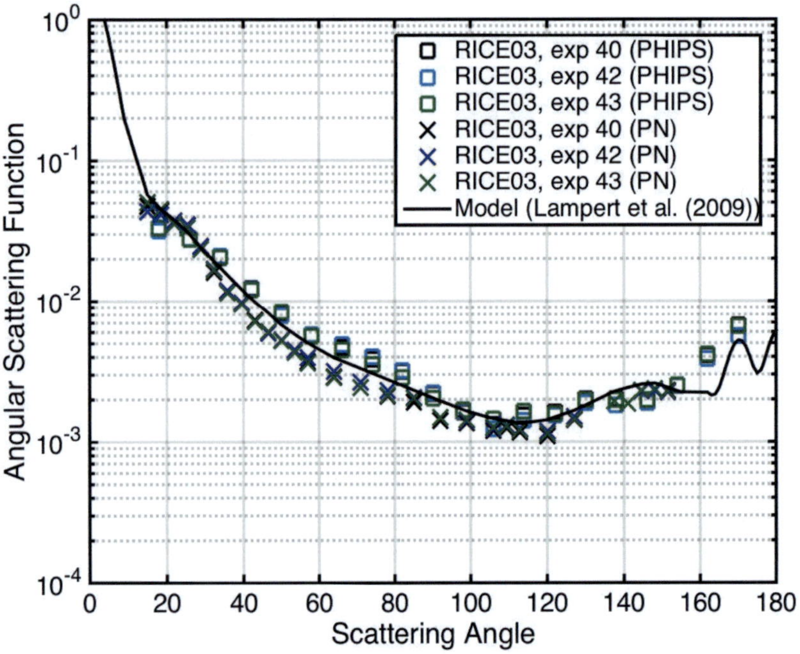

Figure 5.9.: Angular scattering functions of hexagonal columns measured with the PHIPS-HALO (squares) and the PN (crosses) in different experiments.

The second feature that we see in Fig. 5.9 is that there is almost no difference between the PN measurements and the PHIPS-HALO measurements. Some systematic differences are observed, but these can be instrument specific or are due to the difference in the laser wavelengths (PHIPS-HALO 532 nm and PN 785 nm). This is the first instrument comparison between PHIPS-HALO and a polar nephelometer and is, therefore, an important proof-of-concept for the polar nephelometer part of the PHIPS-HALO instrument.
In the case of near-spherical ice particles we observed a significant difference in the angular scattering function between complex and optically spherical frozen droplets. Fig. 5.8 shows the behaviour of the measured asymmetry parameter, g, during the droplet freezing experiment (Fig. 5.3). A maximum value of 0.85 was measured in the liquid droplet cloud and a minimum value of 0.74 after the complete glaciation of the cloud, when the highest crystal small-scale complexity was measured. These observations are consistent with previous study of Cole et al. (2014), where low asymmetry parameters was detected in the case of roughened cirrus ice particles. In the sublimation period, the g was observed to increase as the complexity of the frozen droplets decreased. Almost the same value for g was reached at the end of the experiment, as was measured for the initial liquid droplet cloud, hence confirming the previous observations that the sublimating frozen droplets can behave optically equivalent to spheres.

Fig. 5.10 highlights the dramatic effect that crystal complexity has on the angular scattering properties of frozen droplets. The angular scattering function of roughened frozen droplets measured in the laboratory simulations (red squares in Fig. 5.10) is smooth, featureless at scattering angles less than 100°, and has an enhanced scattering intensity in the backward hemisphere. The angular scattering response of the rough frozen droplets does not significantly differ from that of roughened columns measured in simulated cirrus clouds (orange squares, see Sec. 5.1.1). Therefore, we can conclude that in these two laboratory cases it is difficult to identify the underlying shape of the ice particles from the angular scattering function, but instead the crystal complexity seems to dominate the scattering properties.

After the roughness layer had been sublimated from the surface of the frozen droplets, they scattered light similar to water droplets (dark blue squares in Fig. 5.10) and show droplet-like features, i.e. minimum between 100 and 120° and an rainbow-like feature (in this case ice bow) around 140°. Similarly, Gayet et al. (2012a) observed an ice-bow like feature around the same angle for near-spherical ice particles at the top of a convective storm. Baran et al. (2012b) was able to explain this feature by assuming the additional presence of quasi-spherical ice particles. The ice bow-like feature is also observed in the case of the roughened frozen droplets, however, having a peak around 130°. This shift in the ice bow can be modelled by increasing the distortion of the quasi-spherical model particles (Baran et al., 2012b). Baran et al. (2012b) also argued that the underlying spherical shape of the ice particles can survive the addition of surface roughness or distortion. However, an ice bow was also observed in the case of roughened or complex columns (orange squares in Fig. 5.10), which could indicate that it might be a more universal feature in ice clouds and not only related to spherical ice particles.

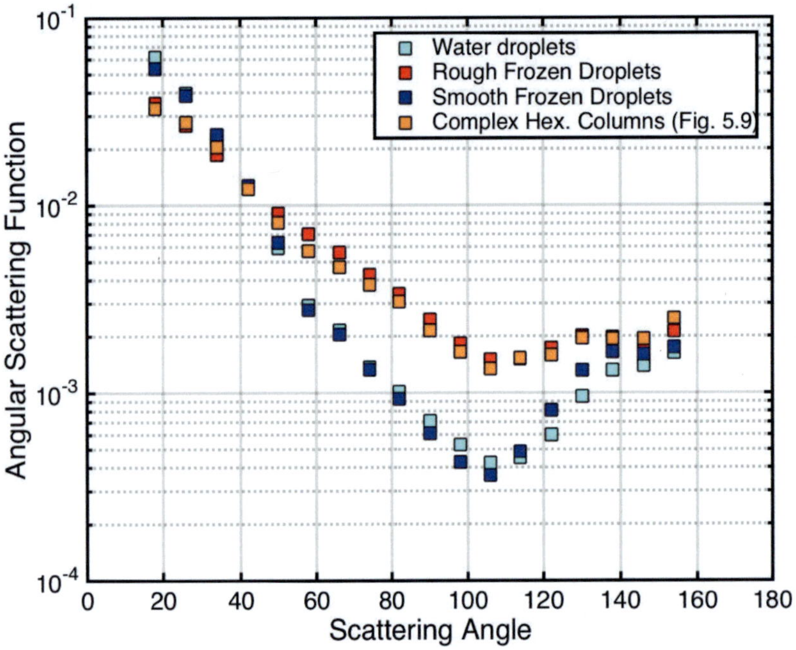

Figure 5.10.: Measured angular scattering functions for water droplets (light blue squares), frozen droplets during their growth (red squares), optically spherical frozen droplets at the end of the sublimation period (dark blue squares), and for roughened columns at -50°C (orange squares; see Sec. 5.1.1)

5.4. The influence of ice crystal complexity on the depolarisation properties of ice clouds

Modelling results reveal substantial differences in the angular depolarisation and polarisation properties of complex ice crystals compared to their pristine versions (Takano and Liou, 1995; Baum et al., 2010). The most sensitive angular region for identifying complexity is normally between 120° and 170°, whereas at exact backward direction (180°) the differences in the depolarisation properties can be smaller (Baum et al., 2010). Specifically, the information of the backscattering depolarisation properties of smooth and complex ice particles is important for interpreting depolarisation lidar observations, like those made with the CALIOP[1] instrument on the CALIPSO[2] satellite platform.

Here, we discuss the linear and circular depolarisation ratio measured in the case of hexagonal columns (Sec. 5.1.1) and near-spherical ice particles (Sec. 5.1.2). From the previous chapter we learned that the depolarisation properties of both dust aerosol particles and SOA particles were strongly size-dependent. Therefore, it is not surprising that we see a strong size dependence also in the case of hexagonal ice particles (Fig. 5.11). The depolarisation ratios of the sub-100 µm hexagonal columns varied between 0.12 to 0.36. The minimum in depolarisation ratio was observed for particles with a GMD of 10 µm and the depolarisation ratio increased sharply to smaller and larger sizes. The measurements were done at two different starting temperatures: 233 K (blue dots) and 223 K (red dots). We see that the curve in case of 223 K experiments is shifted somewhat to smaller sizes. Yet, it seems that the temperature has only negligible effect to the depolarisation ratio compared to the effect of the size. In general, the observed depolarisation ratios of small, <100 µm, hexagonal ice particles were comparable to the atmospheric polarisation lidar measurements (Sassen, 1991; Chen et al., 2002; Noel et al., 2004) and to other laboratory measurements (Sassen, 1991; Smith et al., 2016).

As explained in Sec. 5.1.1 the hexagonal columns were produced with different degrees of complexity. I investigated the effect of the particle complexity on their depolarisation properties in a narrow size range to avoid bias due to the size-dependency. Fig. 5.12 shows the linear and circular depolarisation ratio of the hexagonal columns that had the GMD between 9 µm and 10 µm in the regrowth period. In this size range the linear depolarisation ratio was the

[1] Cloud-Aerosol Lidar with Orthogonal Polarisation
[2] Cloud-Aerosol Lidar and Infrared Pathfinder Satellite Observation

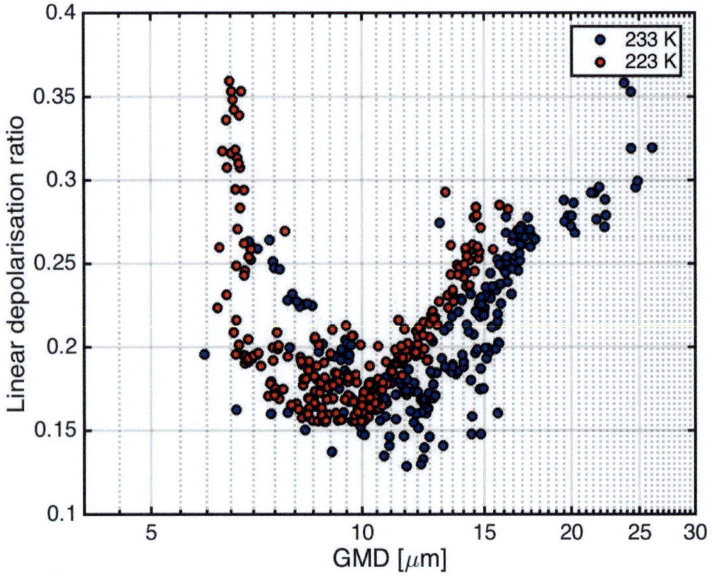

Figure 5.11.: The linear depolarisation ratio as a function of GMD (from the PPD-2K) for hexagonal ice particles nucleated around 233 K (blue dots) and around 223 K (red dots).

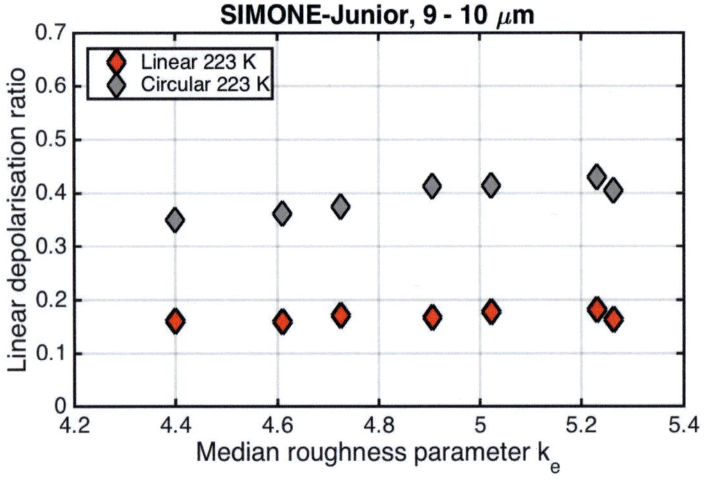

Figure 5.12.: The linear depolarisation ratio as a function of particle complexity for hexagonal ice particles nucleated around 233 K.

most stable (Fig. 5.11). We detect almost no change in the linear depolarisation ratio with varying complexity parameter from 4.40 to 5.26. A similar conclusion was made also for other size ranges. In the circular depolarisation ratio we see a slight increase in the depolarisation ratio with increasing complexity. This slight change in the depolarisation properties in case of circular incident polarisation illustrates the higher sensitivity of the circular depolarisation ratio to detect changes in the microphysical structure of the particles (Mishchenko and Hovenier, 1995b).

The slight increase in depolarisation ratio is in contrast to modelling studies that predict a decrease in the depolarisation properties with increasing complexity (Takano and Liou, 1995). Furthermore, the modelling studies in case of columns predict larger differences in the depolarisation properties. It is not yet fully clear, why we do not see a significant change in the depolarisation ratios with crystal complexity. One reason could be that we measure too close to the exact backward direction, where the differences in the depolarisation properties can be smaller. Therefore, measurements of depolarisation properties at a larger offset from the exact backward direction might reveal more details of the particle morphology.

Unlike the observations made with hexagonal columns, the near-spherical frozen droplets showed a distinct dependency between particle complexity and their depolarisation ratio. In Fig. 5.3 we see the development of the linear depolarisation ratio in a simulation experiment. The maximum measured linear depolarisation ratio was around 0.3 in the case of complex frozen droplets with a GMD around 20 µm. This roughly corresponds to the depolarisation ratio measured for 20 µm hexagonal columns (Fig. 5.11). As the complexity was sublimated from the frozen droplets the depolarisation ratio decreased and a depolarisation ratio of 0.1 was measured before the particles were diluted. This decrease is partly due to the reduction in the particle size, as seen in the case of hexagonal particles. However, unlike with hexagonal particles, the increase in the depolarisation at sub-10 µm sizes is missing. Instead the depolarisation values stay low. Therefore, in the sublimation period, where the particle complexity is below the defined threshold, the particle shape is dominating the depolarisation properties of the frozen droplets. Similarly, in the thunderstorm simulations with the small and near-spherical frozen droplets (Figs. 5.7a and 5.7b) we observed a low and almost constant depolarisation ratio around 0.05–0.1.

5.5. Summary and conclusions

In this chapter we learned that the ice particle growth conditions, i.e. the available condensable water vapour (ζ_v^{acw}), govern the degree of ice particle small-scale complexity. At cold cirrus temperatures (<-40°C) the ice particles are formed and grown through the vapour phase, which results in hexagonal columnar ice particles. For these hexagonal particles, a strong randomisation of the light scattering properties was found as the consequence of their surface complexity. This resulted in featureless cloud angular scattering functions with a stable asymmetry parameter g that was insensitive to further changes in the growth conditions.

Similar angular scattering functions were measured for near-spherical but complex ice particles, indicating that crystal complexity could remove the shape information from the angular scattering function and, therefore, dominate the angular light scattering of ice particles. However, it was shown that if the small-scale complexity was disposed from the near-spherical ice particles by sublimation, the resulting ice particles showed an angular light scattering behaviour similar to liquid droplets, therefore, again revealing the shape-dependent features. Indeed, a maximum difference in the angular scattering function and the asymmetry parameter was observed just by changing the surface properties of these particles.

In contrast to the angular scattering function, the depolarisation properties of the ice particles were found to be rather insensitive to changes in crystal complexity. Only in the special case of near-spherical ice particles, the sublimation of the crystal complexity resulted in a clearly measurable change in the depolarisation properties. This, again, highlights the sensitivity of the depolarisation measurements to deviations from a spherical shape, as was observed in the case of viscous α-pinene particles.

6. The degree of complexity of atmospheric ice particles and measurements of their angular light scattering

It was shown in the laboratory experiments that a not too high supersaturation or excess water vapour mass is needed to induce ice particle small-scale complexity that result in a stable angular light scattering function. However, for the radiative forcing it is important to know, whether real atmospheric ice particles can be assumed to be complex in general. To answer this question, I applied the methods developed in the previous chapters to quantify the ice particle complexity in natural cirrus clouds and its influence on the angular light scattering function. I operated the SID-3/PHIPS-HALO instrument package onboard the HALO research aircraft during the ACRIDICON-CHUVA and ML-CIRRUS campaigns. In these campaigns, I gathered a comprehensive dataset of tropical and mid-latitude cirrus ice particle properties that I later expanded by analysing SID-3 measurements conducted onboard the WB-57 and Polar-6 aircrafts during the MACPEX and RACEPAC field campaigns, respectively. This chapter begins with discussing the results from selected missions from the ACRIDICON-CHUVA and ML-CIRRUS campaigns. At the end of this chapter, I summarise a collective SID-3 dataset from the all aircraft missions, where SID-3 was operational. This covers ice particle complexity from the tropics over mid-latitudes up to the arctic.

6.1. Small- and large-scale ice crystal complexity in tropical cirrus during the ACRIDICON-CHUVA campaign

There are some significant differences between the ice particles found in the tropics and in mid-latitude regions. In the tropics, deep convection drives aerosol particles and humidity from the boundary layer towards the upper troposphere and in these vigorous systems ice particles are formed and evolved by a variety of processes (e.g. homogeneous and heterogeneous nucleation, depositional growth and riming, sublimation, etc.). The ice particles can be transported from the convective cell in so-called anvil outflows that can spread over large areas and lead to a very high cirrus cloud coverage in the tropics, up to 70% (Nazaryan et al., 2008). Therefore, the tropics have traditionally been a popular target for dedicated aircraft missions (e.g. Stith et al., 2002; Lawson et al., 2003; Stith et al., 2004; Connolly et al., 2005; Heymsfield et al., 2005; May et al., 2008; Lawson et al., 2010; Frey et al., 2011).

This tradition was continued in September-October 2014, when the HALO aircraft was based in Manaus as a part of the German-Brazilian ACRIDICON-CHUVA campaign (Wendisch et al., 2016). The goal of this campaign was to quantify aerosol-cloud-precipitation interactions and their thermodynamic, dynamic, and radiative effects by in-situ and remote sensing measurements over Amazonia. During the ACRIDICON part of the campaign, HALO performed 14 research flights to target five scientific topics: (a) cloud vertical evolution and life cycle (cloud profiling), (b) cloud processing of aerosol particles and trace gases (inflow and outflow), (c) satellite and radar validation (cloud products), (d) vertical transport and mixing (tracer experiment), and (e) cloud formation over forested/deforested areas. Flights were conducted at polluted and near-pristine environments. The near-pristine flights were conducted north of Manaus, where there is very little biomass burning, or over the Atlantic ocean. The polluted flights targeted the Manaus plume and the cloud systems south of Manaus, where frequent biomass burning is a source of pollutants that can be transported to the upper troposphere in convective systems.

Figure 6.1.: The SID-3 instrument (instrument with white cover installed on the centre pylon) onboard HALO over the Amazonian rainforest.

6.1.1. Selected flight missions

From the 14 research flights I selected five flights to be discussed in this thesis. The criteria for selecting the flights was that they either were outflow missions or that cirrus clouds were sampled with good statistics as a part of another mission. In four of these flights (AC11, AC15, AC16, and AC20) tropical anvil outflows were systematically studied at different heights and/or distances to the core. For a comparison, I chose a flight, where in-situ formed cirrus was sampled (AC12). Figure 6.2 shows the trajectories of all of these five flights. Two of the outflow flights (AC15 and AC20) probed convective systems south of Manaus, whereas in AC11 outflows north of Manaus and in AC16 west of Manaus were studied. The convective systems south of Manaus can be considered to be more polluted compared to the convective system north of Manaus based on the expected increase in anthropogenic emissions due to biomass burning. I quantified the pollution level for each flight by determining the aerosol concentration at the boundary layer. For this, I used the CPC onboard HALO operated by the group of Prof. B. Weinzierl from DLR, Oberpfaffenhofen. Table 6.1 summarises the boundary layer aerosol concentrations for each outflow flight. The convective system north of Manaus was clearly the most clean of all the cases. However, it should be noted that an aerosol concentration of nearly $500\,\mathrm{cm}^{-3}$ cannot be classified as pristine environment. In that year the start of the rain season was delayed, so finding unpolluted or pristine conditions was almost impossible. In contrast to the AC11 flight, the flight AC15 was intermediate polluted with a boundary layer aerosol concentration of $1100\,\mathrm{cm}^{-3}$ and the later flights AC16 and AC20 were the most polluted with boundary layer concentrations of $2000\,\mathrm{cm}^{-3}$. The effects of pollution to the ice particle microphysics is discussed in the detailed descriptions of the selected flight missions.

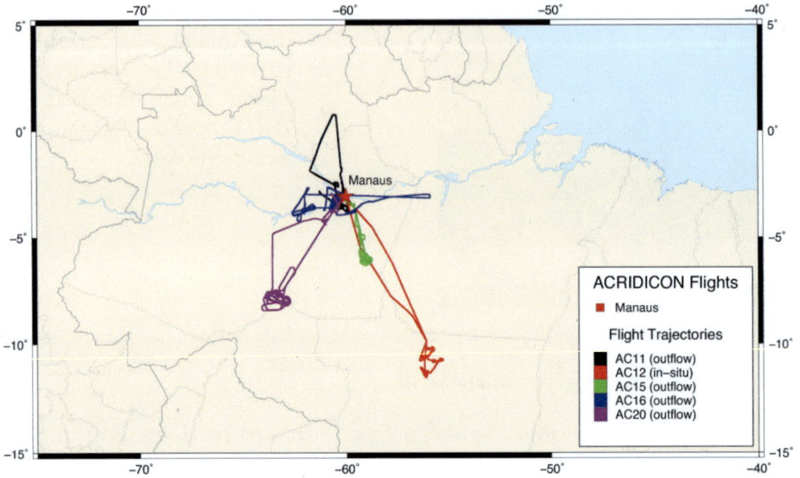

Figure 6.2.: Flight trajectories of the selected missions from ACRIDICON-CHUVA.

Flight	Aerosol Conc. [cm^{-3}]
AC11	490
AC15	1100
AC16	2000
AC20	2000

Table 6.1.: The boundary layer aerosol concentrations for the selected flights.

AC11

On 16 September 2014 developing young convective systems were searched north of Manaus as a part of an inflow and outflow mission. The mission PI for the flight was Dr. H. Schlager. With the help of satellite images the ground crew detected developing convective systems northwest of Manaus, and the HALO aircraft reached the area about 1.5 hours after their formation. Fig. 6.3a shows the flight path through the convective outflow at an altitude of 12.7 km, where the air temperature was measured to be −53°C. The HALO aircraft traversed two separate outflows from north to south. A photograph of the second system was taken after the HALO exited the first system (Fig. 6.3b).

Visual inspection of the PHIPS-HALO ice particle images revealed a significant amount of small (<50 µm) ice particles (Fig. 6.4). Overall 23% of the imaged ice particles were classified as small frozen droplets and 19% as other small irregular ice particles. With smaller fractions plates (9%), bullet rosettes (14%), columns (3%) and aggregated ice particles (15%) were observed.

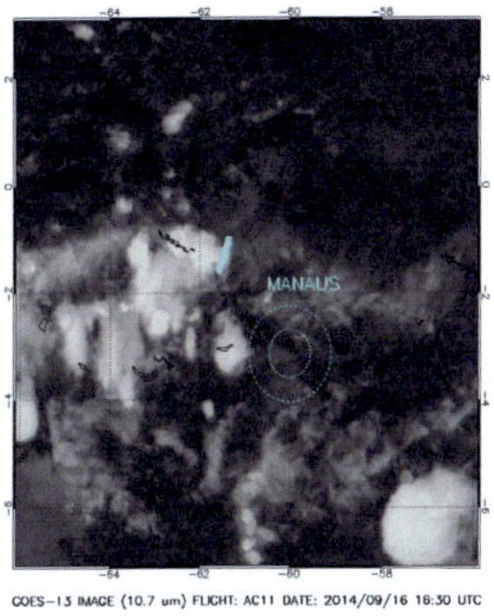

(a) HALO flightpath from flight AC11.

(b) Photograph of the second outflow. Courtesy of Hans Schlager.

Figure 6.3.: First outflow sampling during AC11.

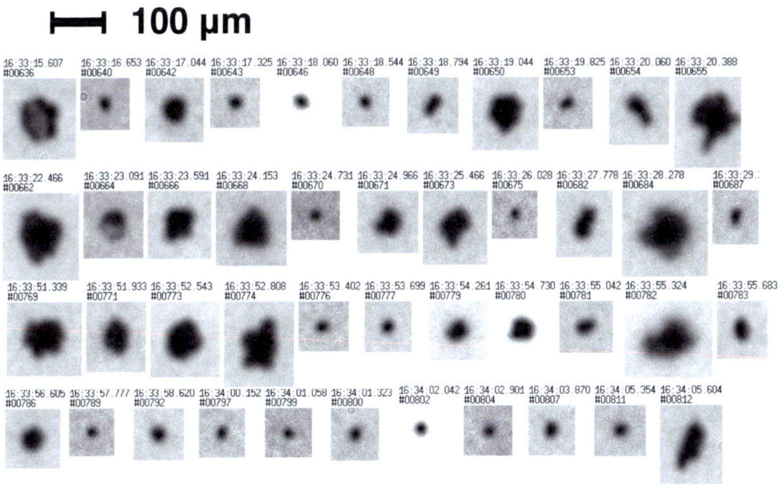

Figure 6.4.: PHIPS-HALO images of ice particles during AC11.

Later on the same flight the HALO aircraft observed a second young developing convective system. The system was reached at 18:28 UTC and sampled at three increasing altitudes. The temperatures during the different sampling altitudes were −33°C, −44°C and −48°C, respectively. At the lowest altitude

I observed traces of supercooled liquid water, but too few ice particles for a statistical analysis. At the highest altitudes below the homogeneous freezing threshold, the ice particle habits were dominated by small ice particles, similar to the first sampled outflow. The fraction of small particles increased from 42% to 70% as the highest altitude was reached. Other habits were plates (30% at lower altitudes and 15% at higher altitudes) and aggregates of plates (11% at all altitudes).

The clear dominance of small ice particles and frozen droplets in the AC11 outflows could be explained with several factors. First, very little ice was observed in the lowest flight level at −33°C, which would suggest that the freezing level was shifted to relatively cold temperatures. At colder temperatures, the more ice particles can be formed simultaneously, which increases the total ice surface and hinders the growth of large ice crystals. Secondly, the boundary layer aerosol concentration was the lowest on that day and in that area, which could be a sign of an aerosol effect, or to be more explicit, the lack of them. Lack of good heterogeneous ice nuclei could promote homogeneous freezing, which in turn would lead to rapid glaciation and formation of small near-spherical frozen droplets (Sec. 5.1.2). Such small and near-spherical ice particles can be observed from the PHIPS-HALO images (Fig. 6.4). Nevertheless, it was apparent that the sampled outflows on that day did not represent the typical tropical anvil ice microphysics (Stith et al., 2002; Lawson et al., 2003; Stith et al., 2004; Connolly et al., 2005; Heymsfield et al., 2005; May et al., 2008; Lawson et al., 2010; Frey et al., 2011), but resembled more of a typical mid-latitude convective system (Stith et al., 2014).

AC12

On 18 September 2014 the HALO aircraft headed for a cloud profiling mission south of Manaus led by Prof. M. O. Andreae. Not long after take-off, HALO encountered a cirrus cloud at an altitude of 12 km and a temperature of −49°C. The cirrus cloud was sampled at a constant altitude and the duration of the cirrus flight leg lasted around 10 minutes so that enough statistics could be gathered. Inspection of the ice particle images showed that the dominating ice particle habits were small columnar ice particles (56%) and larger bullet rosettes (34%), ice particle habits typical for in-situ formed cirrus (Fig. 6.5). Furthermore, convective cells were absent in that region, so the cloud could be classified as an in-situ formed cirrus.

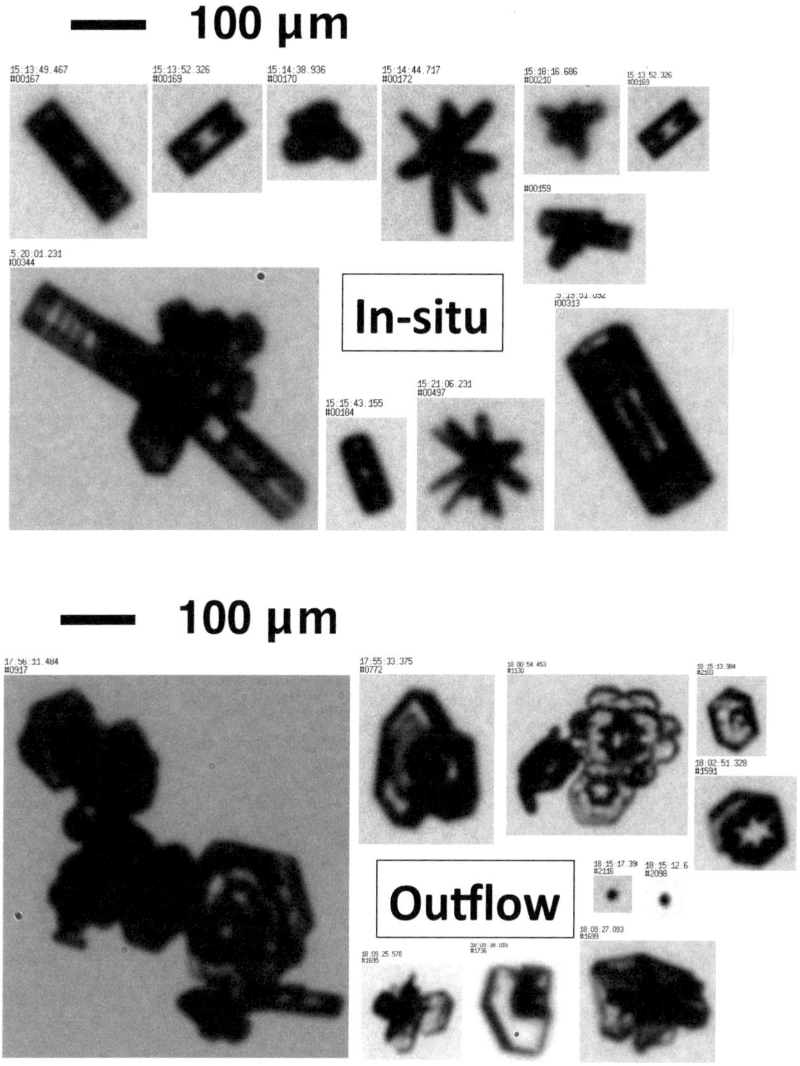

Figure 6.5.: Examples of ice particles imaged by the PHIPS-HALO in an in-situ cirrus (AC12) and in an outflow cirrus (AC20).

AC15

On 23 September a radiation and satellite validation flight was performed south of Manaus led by Prof. M. Wendisch. At the end of the flight a cumulonimbus system with a heavy outflow was observed and, although the flight was not designed as an outflow mission, the outflow was decided to be sampled systematically at different heights. Altogether eight flight penetrations with circular flight patterns were made at decreasing heights from 13.8 km to 11.3 km corresponding to temperatures from −65°C to −44°C. Therefore, AC15 was the first mission, where an outflow was sampled height-segregated with a high resolution, so that an accurate image of the vertical structure of the outflow could be obtained.

Figure 6.6 shows the habit fractions derived from the PHIPS-HALO instrument through a manual classification. Here, for the sake of simplicity I categorised the different habits into four categories: plates, aggregates of plates, small ice particles and columns. I omitted bullet rosettes and other aggregated habits because their fractions were negligible. Small ice particles comprehend everything that have a diameter smaller than 50 μm including frozen droplets but don't show a clear columnar shape according to the instrument resolution. Columns were classified as a separate class because they can be only formed at the cirrus temperatures and, thus, can be used to investigate ice particle formation after mixed-phase conditions.

Similar to AC11 we observe a high fraction of small ice particles (from 28% to 43%). However, unlike in AC11 we observe now a larger fraction of plate-like particles. Aggregated plates are observed almost with the same concentration as the small ice particles (from 20% to 42%) and single plates with a somewhat smaller fraction around 20%. The fraction of columns is insignificant, around 3%. The high fraction of plate-like particles indicate that a significant fraction of the ice particles met in the cold outflow were formed in the lower, warmer mixed-phase region and were transported by convection to higher altitudes. This could be influenced by the fact that we also observe now more aerosol particles in the boundary layer that are transported with the convection to higher altitudes and can contribute to ice formation as heterogeneous ice nuclei. However, also other factors, like cloud base temperature and updraft speed, influence the ice formation and icing levels of these systems. It is long known that tropical convective systems are dominated by plate-like particles independent of the location (e.g. Stith et al., 2002; Connolly et al., 2005; Heymsfield et al., 2005; May et al., 2008; Lawson et al., 2010; Frey et al., 2011), which may also be influenced by the relatively warm cloud base temperatures that are typical for the tropics (Paul Lawson, 2015, personal communication).

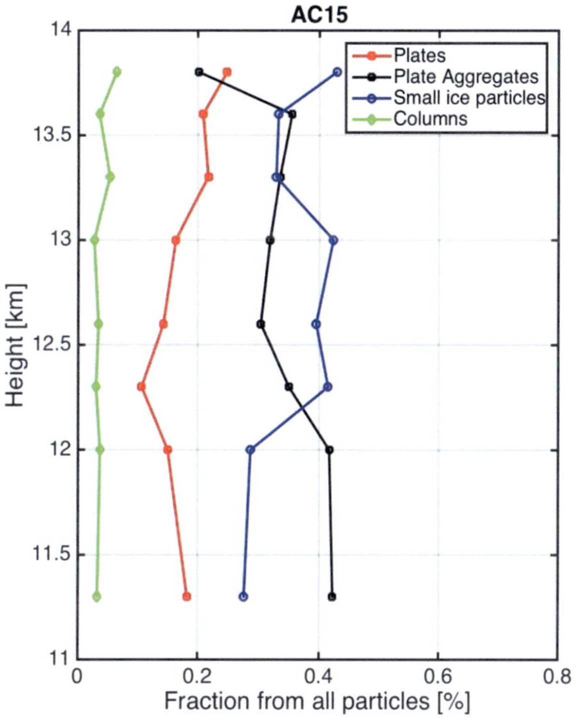

Figure 6.6.: The AC15 habit fractions derived from the PHIPS-HALO through manual classification as a function of height.

AC16

On 25 September 2014 an inflow and outflow mission was performed to the west of Manaus led by Dr. H. Schlager. The mission was started by releasing tracer gas from Manaus and the transportation of the tracer was followed to the west of Manaus. After a two and half hours flight, HALO reached an area with large fires that also resulted in pyro-convection. In this area a large cumulonimbus system was observed and sampled between 17:38-18:34 UTC. Penetrations were made at four different heights corresponding to temperatures from −33°C to −48°C. Additionally, the outflow was sampled at different distances from the core to determine its spatial structure.

Figure 6.7 shows the size distributions of sub-50 µm particles sampled with the SID-3 instrument. The penetrations were made first in the outer part of the anvil outflow and then at the same height closer to the core. As expected, we can see an increase in the particle number concentrations in the inner legs compared to the outer legs. Furthermore, we see that in the three lowest flight altitudes the ice particles are significantly smaller in the outer parts of the anvil than in the inner parts. A possible explanation for this observation is that there exists a distance dependent sedimentation loss of larger ice crystals.

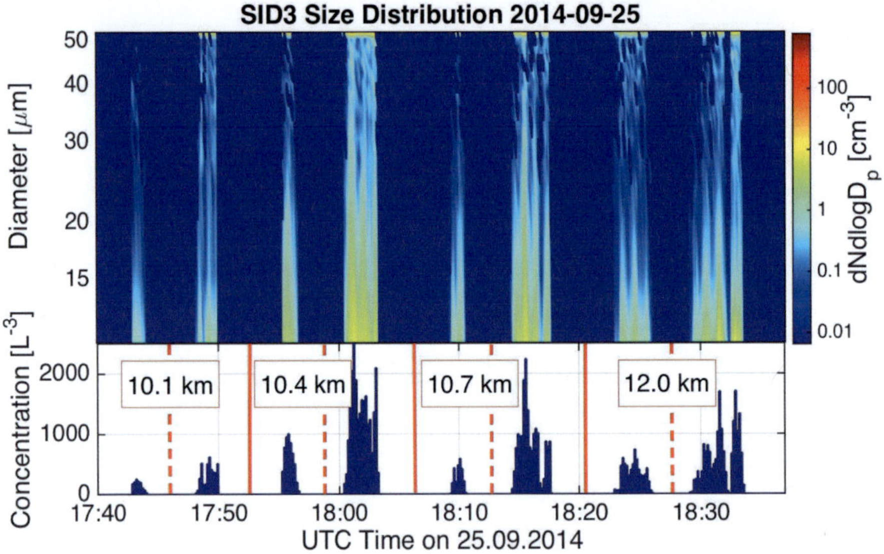

Figure 6.7.: The SID-3 size distribution for flight legs at different altitudes and distances to the core measured during AC16. The different altitudes are separated by solid red lines and the different distances to the core by dashed red lines (in the sequence out-in from left to right).

The investigation of the particle habits show similar habit fractions during AC16 compared to AC15 (Figs. 6.8 and 6.6). At the two highest altitudes the small particle fraction is between 17% and 28%. The fraction of small particles increases towards the lower levels and this increase is more dominant in the outer legs, similar to what was observed in the SID-3 size distributions (Fig. 6.7). In the outer leg at the lowest altitude 53% of the particles are classified as small while at the same altitude but for the inner leg 47% of the particles are small. Also the fractions of plates and plate-aggregates are in general similar to AC15: more aggregated plates are seen (fractions ranging from 30% to 40%) than single plates (fractions ranging from 15% to 24%). However, for the uppermost outer leg the fraction of single plates exceeds the fraction of plate aggregates, which was also observed in the uppermost leg of AC15.

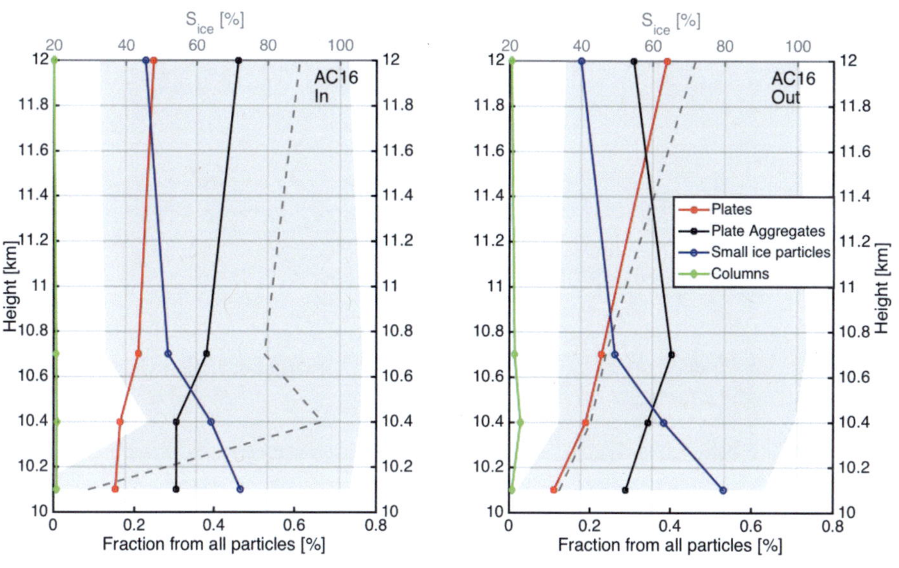

(a) Habit fractions in outer flight legs.

(b) Habit fractions in outer flight legs.

Figure 6.8.: AC16 spatial structure.

Compared to AC15, the AC16 flight shows a more clear spatial structure in the different habits and particle sizes. While during AC15 almost no change in the vertical structure was observed, during AC16 we saw a clear increase of small particle towards the lower altitude levels of the outflow. Also, in the outer legs of AC16 we observed a decrease in the fraction of plate aggregates indicating that these crystals have efficiently sedimented from the top outer layers of the outflow system. A similar decrease in the aggregated particle fraction was also observed during AC15 (Fig. 6.6). As said, the reason for the clearer habit separation in AC16 could be that the outflow was more aged and that the outer and lower regions were already sublimating or the larger particle habits had sedimented out. It can also be that the outflow was sampled mainly at the dissipation zone at lower altitudes, as we don't see the clear decrease in aggregated particles at the top of the system. Also, the maximum sampled altitude was lower, only 12 km, compared to the maximum altitude of 13.7 km reached during AC15.

Figure 6.9.: The sampled outflow during AC20. Photograph by Prof. M. Wendisch.

AC20

The last scientific HALO flight of the ACRIDICON-CHUVA campaign was AC20 on 1 October 2014. The mission PI for this mission was Prof. M. Wendisch and one of the key scientific goals was to study an outflow system with different altitudes and distances from the core. The mission headed southwest of Manaus. After two hours in flight, the HALO aircraft reached an area with several cumulonimbus systems with huge outflows and one of these systems was sampled systematically (Fig. 6.9). First, two horizontally rectangular flight patterns were made at an altitude of 13.8 km and a temperature of −62°C to sample the outflow from two distances from the core. Then, the outflow was continued to be sampled at two distances from the core with decreasing height from 13.6 km to 12.3 km corresponding to temperatures from −60°C to −50°C. Lastly, one penetration was done to the flow direction of the outflow at an altitude of 14.5 km to catch the upper parts of the anvil. However, the outflow had descended in the course of sampling so not enough ice particle statistics could be gathered for the analysis at this height.

Similar to AC15 and AC16, the typical ice particle habits were plate-like particles, i.e. single plates and aggregates of plates (Fig. 6.5). However, one major difference was observed: the lack of small particles compared to all the other flights. Figure 6.10 shows the size distributions of AC11 and AC20 measured with the cloud probes SID-3, CDP, CIP and PIP. The data from the CDP, CIP and PIP was provided by C. Mahnke from University of Mainz. We see that during AC20 the largest ice particles reach sizes of almost one centimetre, whereas during AC11 barely one millimetre was reached. Also, the concentration of sub-50 µm particles is enhanced during AC11 compared to AC20. The different lines in Fig. 6.10 represent different flight legs. Although, the size distributions generally stay rather stable from one flight leg to the other, a few legs show enhanced concentration of larger, super-mm particles. A closer inspection revealed that these flight legs were made in the inner parts of the anvil, which supports the earlier findings that the largest, in this case millimetre-sized, ice particles have precipitated from the outer parts of the anvil.

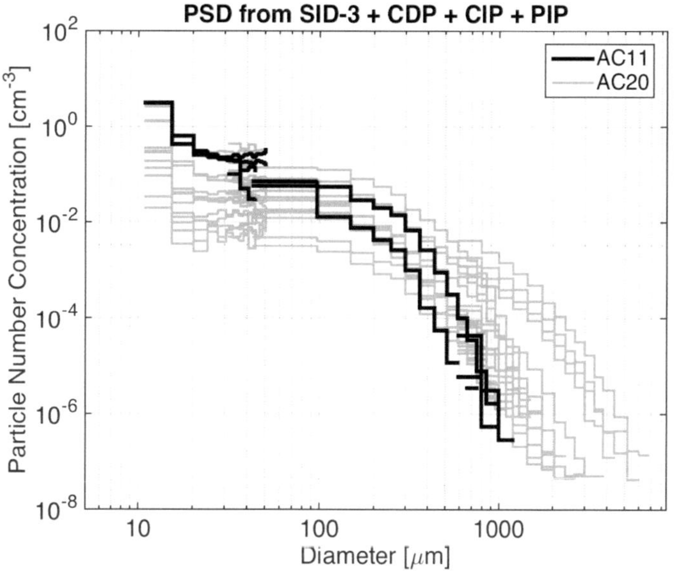

Figure 6.10.: The particle size distribution during AC11 and AC20. The different lines represent the different flight legs.

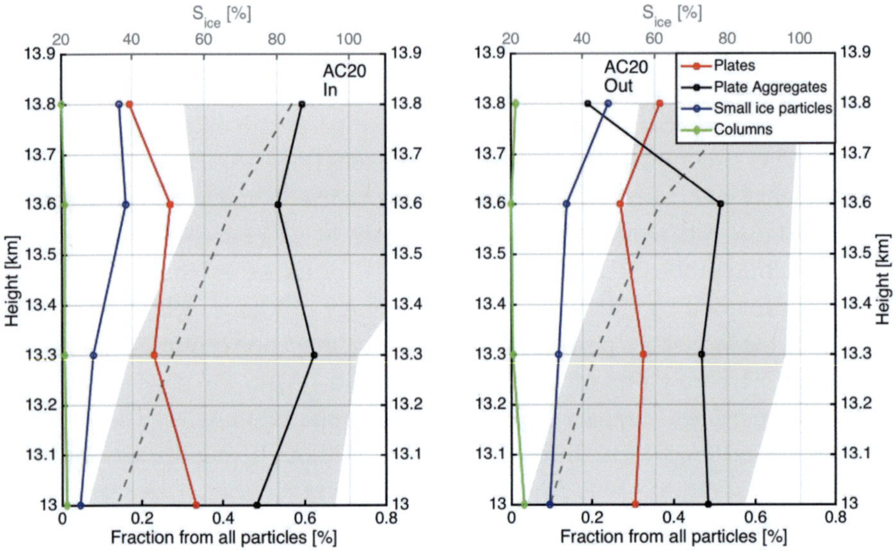

(a) Habit fractions in outer flight legs.

(b) Habit fractions in outer flight legs.

Figure 6.11.: AC20 spatial structure.

I also performed a manual classification of the ice particles habits, similar as it was done for AC15 and AC16 (Fig. 6.11). Since the smaller particles are largely missing, the most dominant particle habit is now the plate-aggregates. Their fractions range from 48% to 62% in the inner parts of the anvil. The fraction of single plates range from 17% to 37% and the small ice particle fraction only from 5% to 15%. Similar to AC15, we do not see a significant variation in the habit fractions at the different heights and now also for the different distances to the core. The only exception again is the uppermost layer at the outer part of the anvil, where the largest aggregates have dissipated and more smaller particles or single habits are observed. Thus, it seems that in general the tropical anvil outflows tend to be well-mixed and homogeneous with respect to particle size and habit up to the upper- and outermost layers. This trend is probably linked to deep convection because we can assume that the higher updraft speeds in those systems might promote better mixing. Therefore, it would be interesting to link the ice particle microphysics in the anvil outflows also to the updraft speeds of the systems. It should be noted that previous studies of tropical convective systems have mainly focused on the phase discrimination in the convective core to determine icing levels or finding precipitation particles. The results presented in this thesis in terms of the spatial structure of convective anvils are, thus, unique and will be an important contribution to understand the cirrus ice microphysical properties in the tropics.

6.1.2. Large-scale complexity

The PHIPS-HALO images (Fig. 6.5) taken from outflow ice particles show complicated shapes, including aggregates and rimed particles. We discussed that in tropical convective systems very high vertical updraft speeds can be achieved, which can enhance the formation of complex habits. The resulting type of crystal complexity can easily be observed with high resolution cloud particle imagers as the scale of the complexity is approximately of the same order as the size of the individual ice crystals. In this thesis such complexity was referred as large-scale complexity.

There are two ways to quantify the large-scale complexity. One can go through every single cloud particle image and look for the presence of aggregation or riming or one can combine the particle dimensional parameters, as deduced from the image analysis, in a power law function to define a complexity parameter. This was briefly explained in Sec. 3.3.3 or is given in details in Schmitt and Heymsfield (2014). The latter method can be conducted automatically, which is of advanced when analysing larger datasets. Here, I applied both methods. I manually classified all the ice particle images from the PHIPS-HALO instrument and compared the manual classification to the derived complexity parameter. This was done, because the complexity parameter can be considered instrument-dependent and in this thesis I investigated the method for the PHIPS-HALO instrument.

Figure 6.12 shows the complexity parameter, C, and the single habit fractions as a function of particle maximum dimensions for the different outflow flights and for the in-situ flight. We observe that aggregated particles are present in each of the outflow flights, whereas during the in-situ flight we observed mainly single habits. We also observe a clear separation in the sizes of the single particles and the aggregated particles: we can roughly classify that particles below $130\,\mu m - 170\,\mu m$ are single habits and particles larger than this are aggregates. Furthermore, the aggregated particles show a strong increase in the complexity parameter as their size increases, which results in a "hockey stick"-kind of scatter plot. The same "hockey stick" patterns can be observed through all of the outflow flights and also in the image analysis from another tropical field campaign, the CRYSTAL-FACE campaign (Schmitt and Heymsfield, 2014).

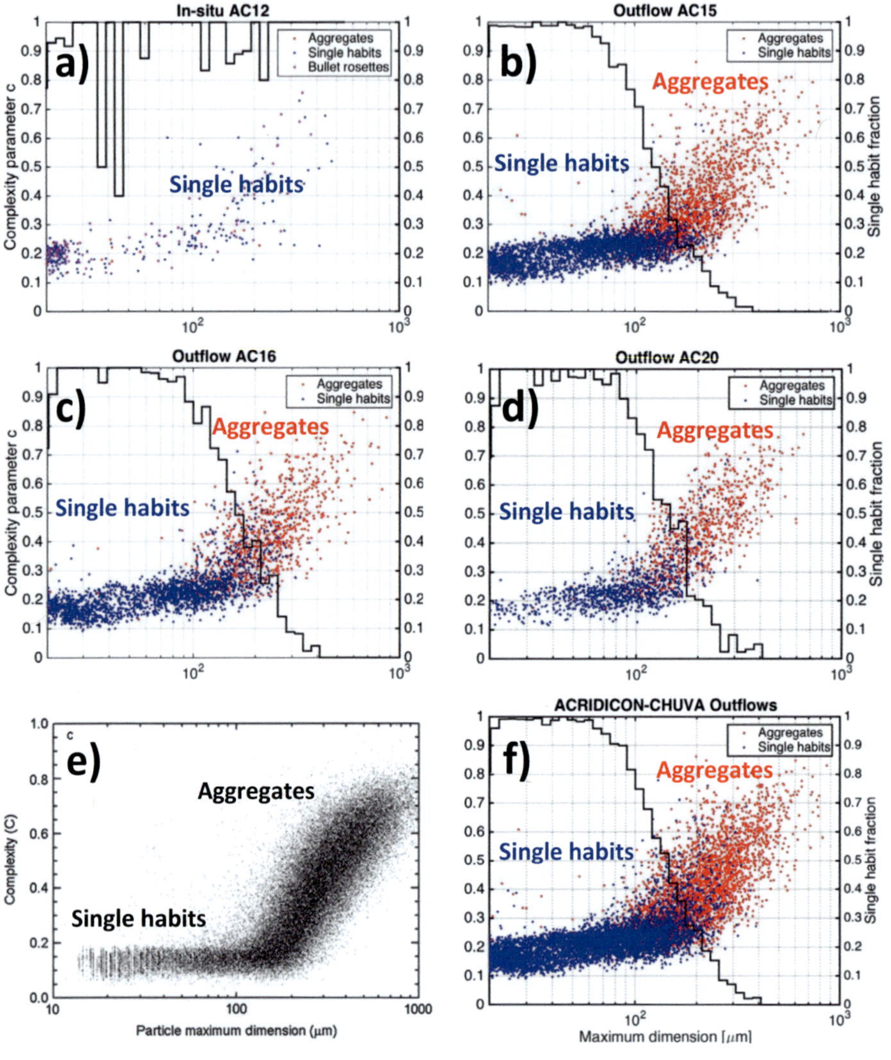

Figure 6.12.: The complexity parameter calculated based on PHIPS-HALO images for (a) AC12 in-situ cirrus, (b) AC15 outflow, (c) AC16 outflow and (d) AC20 outflow. Panel (e) is complexity for tropical CRYSTAL-FACE campaign (Schmitt and Heymsfield, 2014) and panel (f) combines the ACRIDICON-CHUVA results form all outflows.

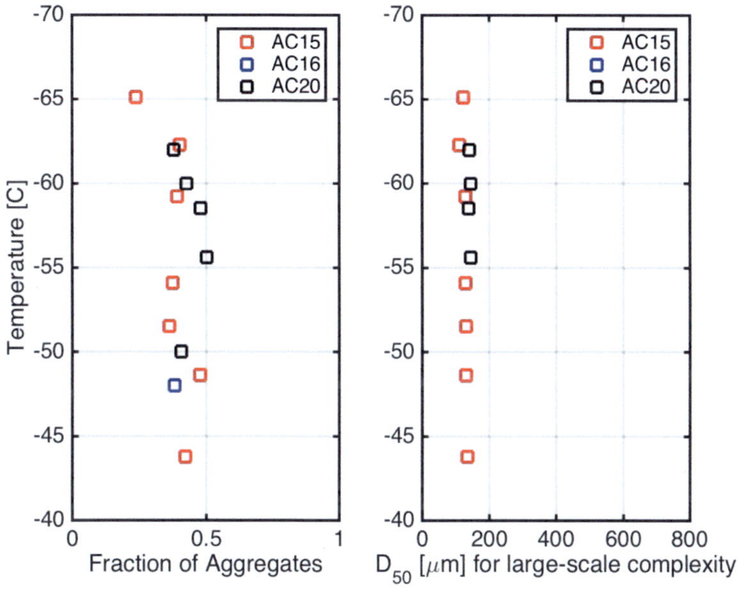

Figure 6.13.: Right: the fraction of aggregates as a function of temperature (in °C). Left: the D_{50} as a function of temperature (in °C).

It seems that the dimensions of aggregates are similar from one outflow to the other. This means that we can easily apply a threshold complexity parameter to separate between single habits and aggregated habits. In the case of the PHIPS-HALO instrument, a threshold complexity parameter of 0.3 is appropriate (cf. the threshold parameter of 0.2 used for the CPI instrument in CRYSTAL-FACE). This threshold value can be used to determine the fraction of aggregates, i.e. the fraction of large-scale complexity. We can also determine the particle dimension D_{50}, where half of the particles are single habits and the other half aggregates. Figure 6.13 shows the fraction of aggregates and the D_{50} as a function of temperature for flights AC15, AC16 and AC20, where the outflows were sampled at different heights. Fig. 6.13 confirms the previous conclusion that the sampled outflows were vertically well-mixed, with the exemption of the uppermost layer. We also see that independent of the height roughly half of the particles imaged by the PHIPS-HALO (>20 μm) are complex, just by considering the large-scale complexity. The sub-170 μm particles are mainly single habits, whose complexity we cannot estimate using the PHIPS-HALO images alone. Luckily, the majority of these particles fall in the measurement range of the SID-3 instrument, so that we can estimate the small-scale complexity of the single habits using the SID-3 method.

6.1.3. Small-scale complexity

The small-scale complexity is governed by the growth conditions of the ice particles. In tropical convective systems we observed that a great fraction of the ice particles were formed though liquid phase in the warmer part of the convection, where we can expect that enough water molecules are present to promote also formation of small-scale complexity. Some evidence of this we can already see form the PHIPS-HALO images of single plates (Fig. 6.5). For example, we observe a lot of cross-plates, where new branches are growing from the basal plane or we see structures inside the plates, e.g. star-like patterns. Pristine plates on the other hand would have unperturbed basal planes that would appear transparent.

To quantify the degree of the small-scale complexity, I analysed the SID-3 scattering patterns using the same method used in the laboratory study. I did a statistical analysis for each outflow, where I combined the different flight legs to get enough statistics. Figure 6.14 shows the results of this statistical analysis for AC11, AC12, AC15, AC16, and AC20. I applied a k_e-threshold of 4.6 to discriminate between pristine and complex particles.

At a first glance the results of this analysis look similar. We have a narrow range of pristine particles and a longer tailing towards increasing small-scale complexity. The fractions of complex particles range from 52% to 82%. The highest degree of small-scale complexity was observed during the AC11 flight (82%). This was also the cleanest flight, where more homogeneous nucleation was to be expected. We learned from the laboratory experiments that homogeneous nucleation can promote a high degree of small-scale complexity, which would explain this increase. The second highest complexity was observed for the in-situ cirrus sampled in AC12 (74%), although we have to take into account that the statistics were relatively low in this case, as only 151 ice particles were accepted in the analysis. Similar degrees of small-scale complexities were observed for AC16 and AC20 (62% and 67%, respectively). In these cases we can expect more heterogeneous nucleation, which could promote the ice particle growth already at lower supersaturations. The lowest degree of small-scale complexity was observed during AC15 (52%), but similar to AC12 the statistics was low, as in this case only 265 ice particles were accepted for the analysis.

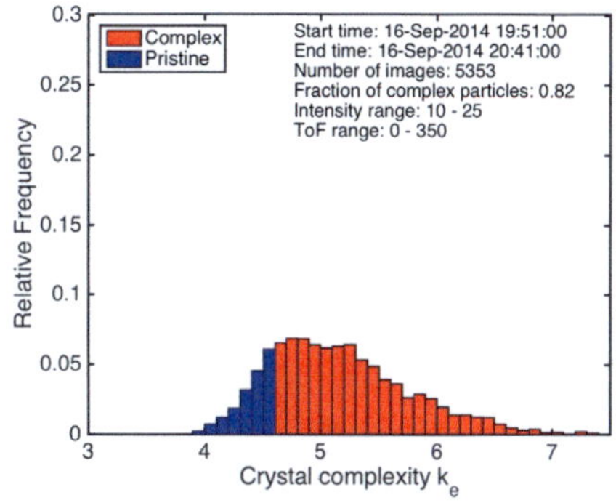

(a) AC11

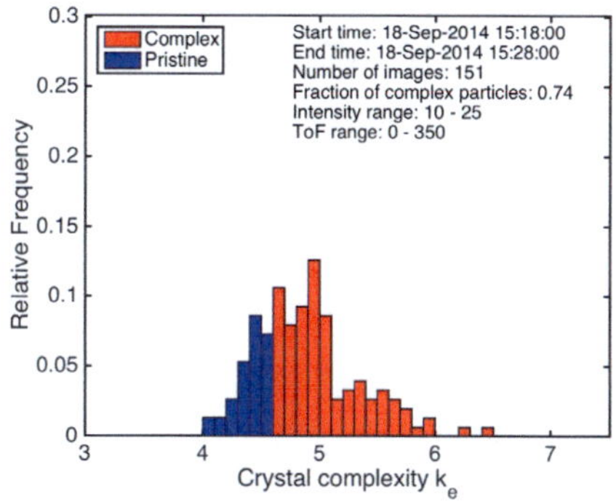

(b) AC12

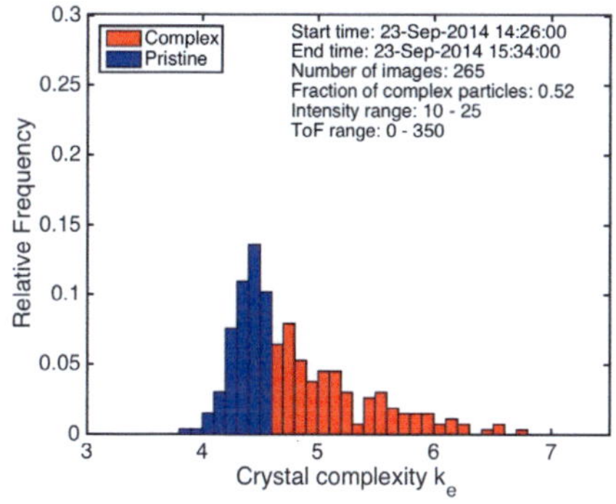

(c) AC15

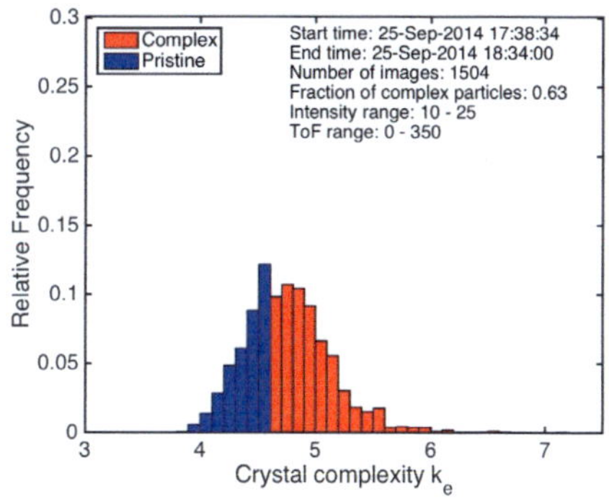

(d) AC16

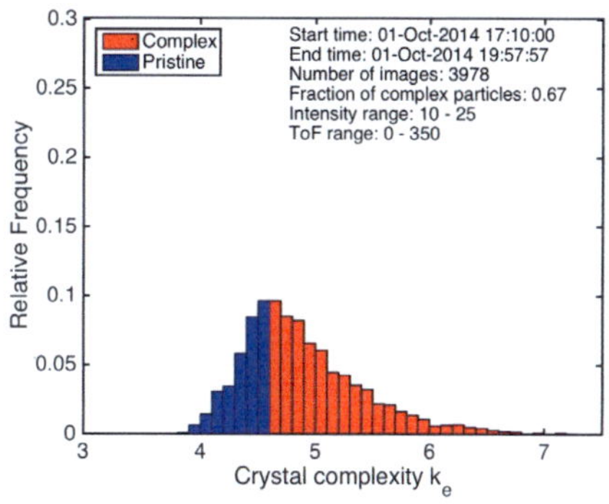

(e) AC20

Figure 6.14.: The statistical analysis of ice crystal small-scale complexity.

In conclusion, we can state that the degree of ice particle complexity measured during the ACRIDICON-CHUVA campaign was rather high, as expected from the laboratory measurements. The median k_e values were 5.1, 4.8, 4.6, 4.7 and 4.8 for the different flights with increasing flight number. This would indicate available condensable water vapour mixing ratios, ζ_v^{acw} (Eq. 5.1), from 5 to 10 ppm according to the laboratory study. As already discussed, the ice particles measured during ACRIDICON-CHUVA were mainly formed in convective systems. In mid-latitudes and in arctic such fast convection is not so common and the cirrus ice particles have different formation pathways. The small-scale complexity of mid-latitude ice particles will be the topic of the next section. Also, it was found in the laboratory study that a k_e value above the threshold of 4.6 leads to uniform angular light scattering; at least in the case of hexagonal ice particles. During ACRIDICON-CHUVA the angular light scattering of atmospheric ice particle could be investigated for the first time with the PHIPS-HALO instrument.

6.1.4. The angular light scattering function of tropical cirrus

I investigated the influence of the observed ice particle complexity on the angular scattering function using the PHIPS-HALO polar nephelometer measurements. These measurements were validated in the AIDA inter-comparison with the established polar nephelometer from the University Blaise Pascal. Fig. 6.15 shows an averaged angular scattering function for each flight and for comparison an averaged angular scattering function from a droplet cloud measured during AC13. For the averaging, I included all the scattering phase functions from all ice particles in every flight leg performed in one cirrus system that were above the background and were not saturated. This roughly corresponds to a size range of 30-300 µm depending on their morphology. Furthermore, all the averaged angular scattering functions are normalised to 1 as described in Sec. 3.3.2.

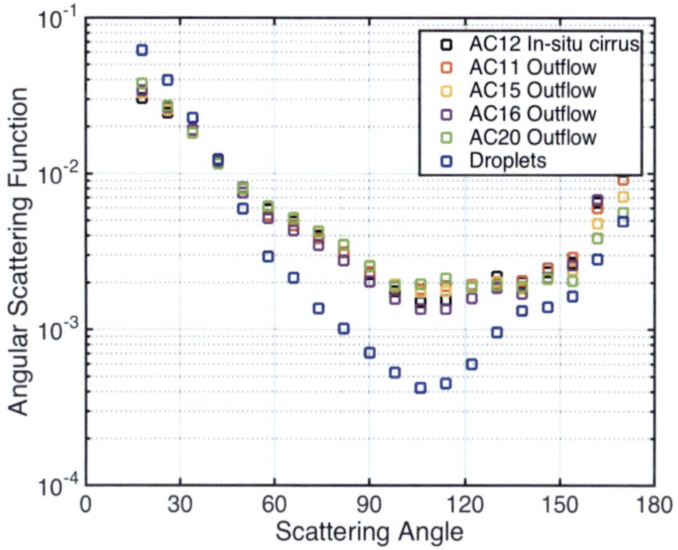

Figure 6.15.: The cloud averaged angular scattering functions for the different flights and for a liquid cloud from AC13.

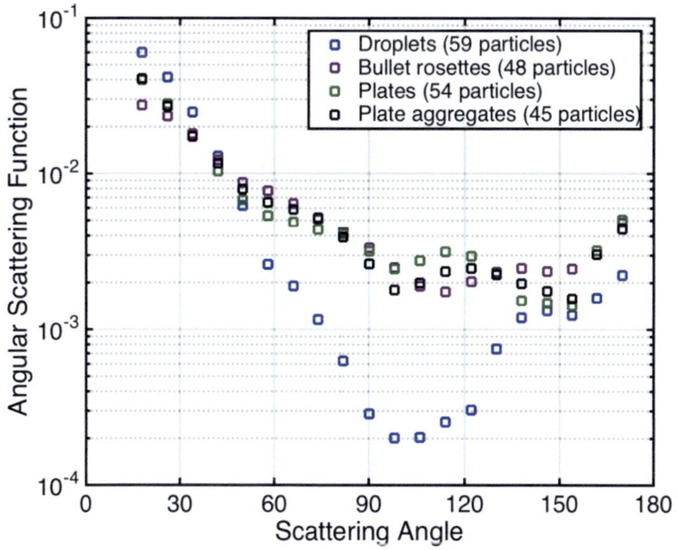

Figure 6.16.: The habit specific angular scattering function for droplets (blue squares), bullet rosettes (purple squares), plates (green squares) and plate aggregates (black squares).

We clearly see that the cirrus angular scattering functions are featureless and have a higher degree of backward scattering compared to the angular scattering function of the liquid cloud. These both are common features for an ensemble ice particle angular scattering. Interestingly, we do not see a significant difference in the angular scattering behaviour between the different cirrus clouds. Now, keep in mind that we observed a different collection of particle habits in the in-situ cirrus compared to the outflow cases (Fig. 6.5). We also observed somewhat different contributions of small ice particles (Figs. 6.10) and saw differences in the small-scale complexity (Fig. 6.14). Yet, the angular scattering function seems to be uninfluenced by these factors. It was already observed in the AIDA experiments that after a certain degree of complexity is reached, the angular scattering function became stable. In all of the ACRIDICON-CHUVA case studies we observed a small-scale complexity above this threshold and also a significant degree of large-scale complexity, all promoting a similar light scattering behaviour. These results together with the AIDA results support the idea of a threshold complexity.

However, as the sampled cirrus clouds consisted always of multiple habits, it is possible that habit specific scattering features are removed already though the mixing of different habits. Now, the advantage of the PHIPS-HALO instrument is that we can link the measured angular scattering function to an image and, therefore identify the particle habit. In the current version of the instrument this assignment can only be done manually. Here, I used the manual image to scattering phase function classification and the performed habit classification to produce averaged angular scattering functions for droplets and three ice habits: bullet rosettes, plates and plate aggregates (Fig. 6.16).

The averaged habit specific angular scattering functions of the ice particles are flat but in comparison to Fig. 6.15 we see somewhat more features. The most distinct features are observed in the case of plates that are the most simple habit, and where we can expect the frequent occurrence of specular reflections. The bullet rosettes and aggregated plates show less features, most likely because their habits are more complex. However, even if the angular scattering functions in Fig. 6.16 are ensemble averaged, we should still be able to see habit specific features, like 22° or 46° halo in the case of single plates. Yet, these features are missing, which can be explained by the small-scale complexity in the case of single habits.

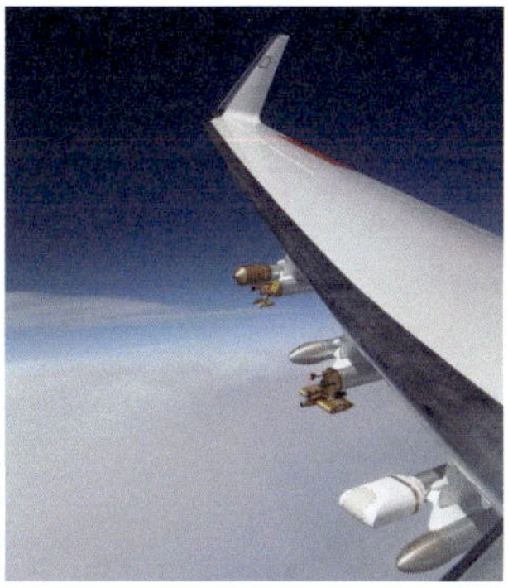

(a) The right wing. The PHIPS-HALO is seen in the uppermost position.

(b) The left wing. The SID-3 is seen in the third position from the cabin.

Figure 6.17.: The underwing instrumentation during ML-CIRRUS

6.2. Ice particle complexity in mid-latitude cirrus

Until now we have discussed the ice particle complexity of natural ice crystals in tropical cirrus, where deep convection is a significant driver for the freezing process. In contrast to the tropical systems, the mid-latitude cirrus systems are typically formed in larger synoptical events, like in warm conveyor belts (WCB), where the updraft speeds are more moderate compared to tropical convective systems. In WCBs the air masses undergo a pressure difference of at least 600 hPa, when they are lifted from near surface to high altitudes, where cirrus formation takes place. During this transport a liquid and/or mixed-phase cloud is formed, so that we can expect that at least a fraction of the cirrus ice particles have a liquid-origin.

A different formation pathway is seen in an in-situ cirrus, where the air masses have not experienced a large vertical transport from the warmer surface but they are formed directly from vapour phase by deposition nucleation or by homogeneous nucleation at cold cirrus temperatures. These type of cirrus clouds are frequently formed along the jet streams that bring westerly air towards Europe. Two separate jet streams influence Europe's meteorological situation. The polar jet stream brings air from northern Canada to northern Europe and the subtropical jet stream from southern USA to Europe's southern coasts.

Furthermore, the northern mid-latitudes are an important area for aviation. The north-atlantic flight corridor is a frequent source of anthropogenic emissions and aviation induced cirrus in the area. They pose an anthropogenic forcing and, therefore, contribute to the uncertainty of cirrus radiative forcing. To better understand the microphysical and radiative properties of aviation induced and natural cirrus clouds in mid-latitudes, a dedicated Mid-Latitude CIRRUS (ML-CIRRUS) campaign took place in spring 2014 over Europe. Altogether 16 research flights were conducted to investigate the natural mid-latitude cirrus systems and anthropogenic aviation induced cirrus, more specific contrails and contrail cirrus. The main goal of the ML-CIRRUS campaign was to study the differences in the cirrus ice microphysics and radiative properties between cirrus clouds of different origin. A manuscript with more information on the HALO instrumentation and the research flights is submitted for publication in the Bulletin of the American Meteorological Society (Voigt et al., 2016).

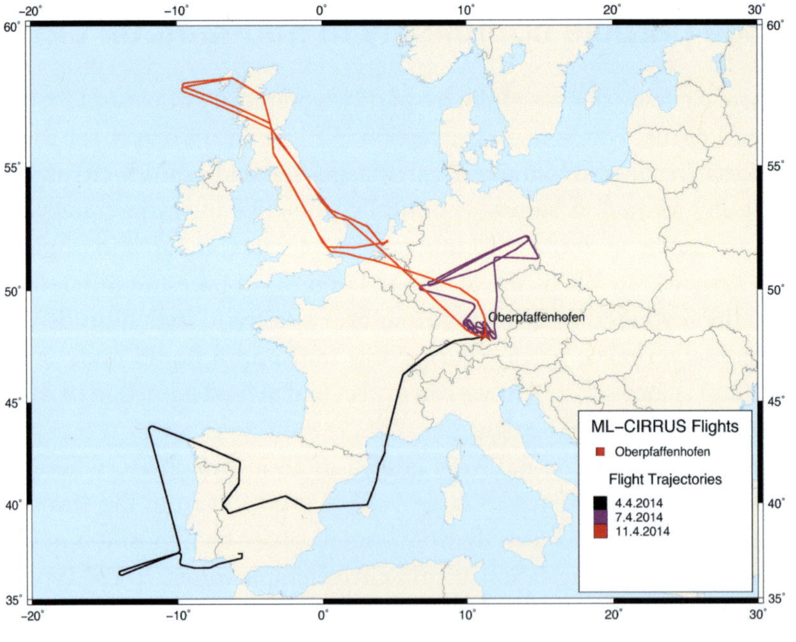

Figure 6.18.: The flight trajectories of three the missions on 4.4.2014, 7.4.2014 and 11.4.2012 performed during ML-CIRRUS and discussed in this thesis.

6.2.1. Selected flight missions over Europe and over continental-USA

Figure 6.18 shows trajectories of the three different ML-CIRRUS flights that are investigated in this thesis. I chose these specific flights because they captured all of the aforementioned cirrus origins: WCB, jet stream formed in-situ cirrus, and contrail cirrus. Additionally, I include one flight from the MACPEX campaign (Spring 2011) that focused on cirrus systems over continental USA. During that flight a convective thunderstorm was sampled and, therefore, this flight is an ideal comparison of the ice particle microphysical properties measured in a mid-latitude convective system compared to the tropical systems measured during ACRIDICON-CHUVA.

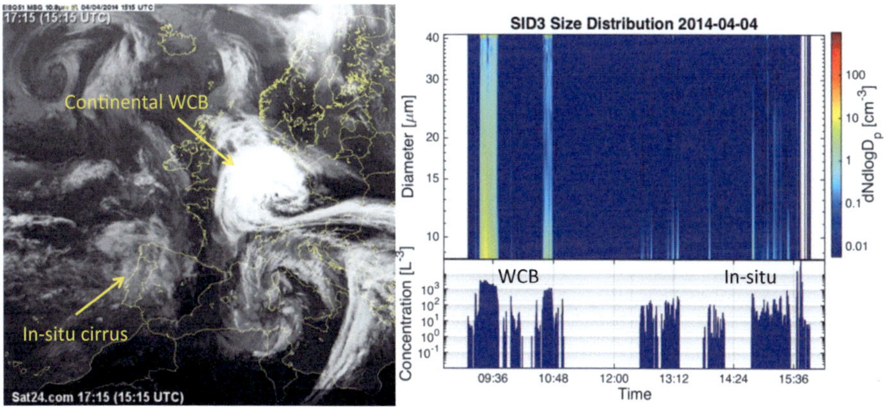

Figure 6.19.: The meteorological situation during the ML-CIRRUS flight on 4 April 2014 and the SID-3 ice particle size distribution measured in this flight. The satellite image is taken by Meteosat in the visible wavelengths.

ML-CIRRUS flight on 4 April 2014

On 4 April 2014 HALO sampled two cirrus systems (Fig. 6.19). The origin of the cirrus systems was defined based on the European Centre for Medium-Range Weather Forecasts (ECMWF) trajectory analysis performed by Prof. H. Wernli from ETH Zürich. Fig. 6.20 shows the results of this analysis for two arrival times on the HALO flight tract. The first time (9:26 UTC) corresponds to the sampling time in the first cirrus system and the second time (15:12 UTC) the sampling time in the second cirrus system. In the first case the trajectories have travelled through a pressure altitude of 800-1000 hPa and experienced a pressure difference of at least 600 hPa, so we can label this system as a WCB. The air masses arrived from north Africa and traveled through Spain before reaching the HALO aircraft and, therefore, we can say that the WCB had a continental origin. Also, around that time there was a Saharan dust outbreak, which affected the sampled air masses by increasing the atmospheric dust load. On the contrary to the first system, the trajectories of the second cirrus system had not spend time below the pressure altitude of 200 hPa for the last 10 days, so we can label this system as an in-situ formed cirrus. The air masses that arrived the HALO aircraft traveled across the Atlantic ocean in the subtropical jet stream, so we can expect relatively low influence by heterogeneous ice nuclei.

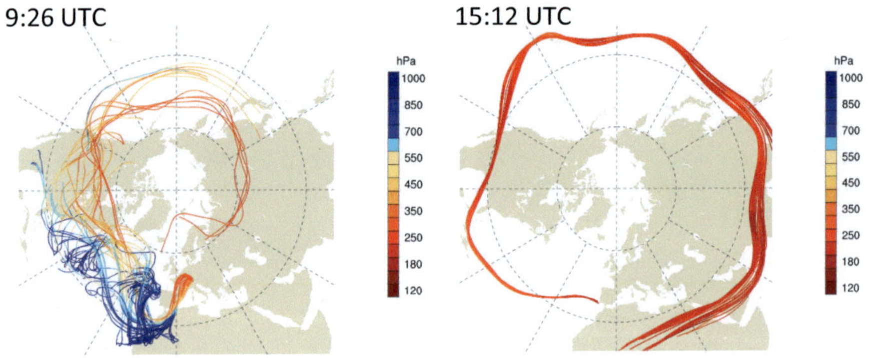

Figure 6.20.: The calculated 10 day backward trajectories for the position of the HALO aircraft at 9:26 UTC and 15:12 UTC. Figures by Prof. H. Wernli.

The continental WCB was sampled at a constant altitude of 9.5 km between 9:15 UTC and 11:00 UTC over France shortly after take-off from Oberpfaffenhofen. After sampling the HALO heated southwest towards Portugal to reach the subtropical jet stream over the Atlantic ocean. However, on that specific day it was not allowed to enter the Portuguese airspace and HALO had to fly around Portugal to access the jet stream (see flight track given in Fig. 6.18). This considerably limited the sampling time in the in-situ cirrus system that had formed west of the coast of Portugal in the subtropical jet stream. HALO reached the in-situ cirrus system at an altitude of 13 km and descended into the cirrus until an altitude of 10.4 km was reached. After this HALO turned around and on the way back accented through the cirrus until a flight level of 12 km was reached.

Fig. 6.19 shows the SID-3 size distribution measured from that flight. The highest ice particle concentrations are seen in the WCB cirrus between 9:15 UTC and 11:00 UTC. Compared to the WCB cirrus, the in-situ cirrus shows clearly lower concentrations between 14:45 UTC and 16:00 UTC. The difference in the particle concentrations can also be observed from the satellite image as a higher optical thickness can be seen in the case of the WCB cirrus compared to the in-situ cirrus. Furthermore, the ice particles in the in-situ cirrus are clearly smaller compared to the WCB cirrus ice particles: the majority is below $40\,\mu m$. The size of cirrus ice particles is probably affected by their origin. The WCB cirrus ice particles were likely grown in mixed-phase conditions, where the Bergeron-Findeisen process can promote the formation of larger ice particles. In the case of jet stream in-situ cirrus, we can expect that the marine origin enhances homogeneous ice nucleation. In such system all ice particles would nucleate almost simultaneously and quickly deplete the available water vapour and, thus, a further growth of the crystals would be prevented.

ML-CIRRUS flight on 7 April 2014

On 7 April 2014 a flight was conducted to target contrails and contrail cirrus over northern Germany (Fig. 6.18). The selected flight route traversed from Frankfurt am Main to Leipzig and back. Over Leipzig it was planned to conduct a lidar overpass flight leg. At that time the area was covered with natural cirrus with a moderate optical thickness and with embedded persistent contrails. Here, I report measurements in two contrails of different ages. The first contrail was sampled on 8:15:22 UTC at an altitude of 11 km and a second contrail was sampled later on 10:23:52 UTC at the same altitude.

Figure 6.21 shows the size distribution measured by the SID-3 during that flight. The HALO aircraft sampled frequently in cirrus clouds, where the observed ice particle sizes were relatively small, below $40\,\mu m$, and the concentrations moderate, $100-200\,l^{-3}$. The short contrail sampling periods are hardly recognisable from the upper figure, so I show the averaged size distributions for the two contrail sampling periods in the lower panel of Figure 6.21. Now we see an increase in the number concentration compared to the natural cirrus: in both contrails a total concentration of $600\,l^{-3}$ was reached. The increase in the particle number concentration is especially seen in the smallest size bin, giving evidence for a rapid formation of micrometer-sized ice particles. Both of these observations are typical for contrail cirrus.

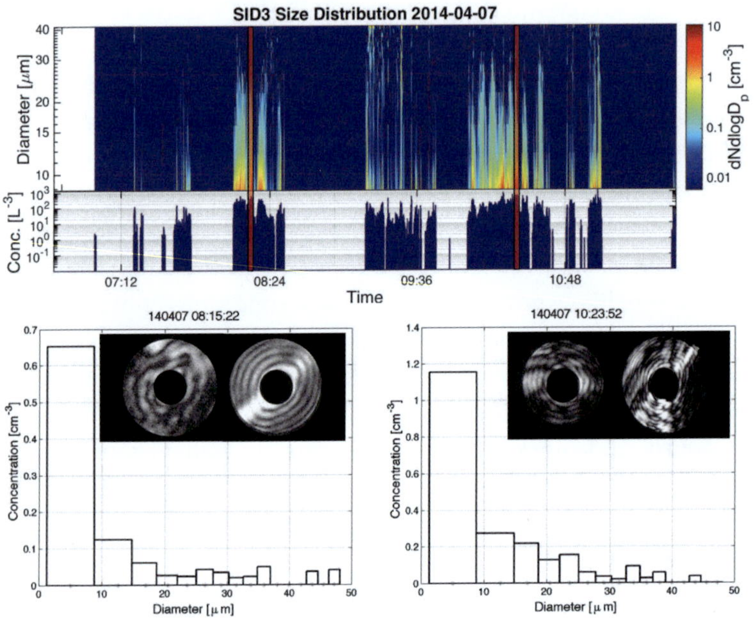

Figure 6.21.: The SID-3 size distribution of the flight on 7 April 2014 over Germany. The highlighted areas in the upper figure mark two contrail sampling periods whose averaged size distributions are shown in the lower panel. The shown SID-3 scattering patterns represent examples of frozen droplets that were smooth enough to still show the concentric ring structure. Such patterns were only occansionally observed in the contrail samples, while the majority of the particles showed a complex scattering pattern.

Contrails are generated, when liquid water droplets form by condensation of water vapour mainly on soot and volatile particles in the exhaust plume (Schumann, 2005; Kärcher and Yu, 2009). In the colder and humid upper troposphere the droplets freeze and form a visible contrail, which spreads out and becomes persistent at ice supersaturated conditions. Therefore, we can expect that the ice microphysical properties in contrail cirrus can differ from the natural cirrus ice particles. Previous studies have shown indications of optically near-spherical ice particles in young contrails. Febvre et al. (2009) found that young (about 2.5 min of age) contrails have a high asymmetry parameter

(0.827) compared to aged (about 20 min of age) contrails (0.787). Similar behaviour was observed in experiments during the CONCERT campaign (Voigt et al., 2010) in an ageing contrail from a A380 aircraft (Gayet et al., 2012b). A decrease in the asymmetry parameter from 0.88 to 0.8 was observed already within the first 5 minutes of contrail evolution. These observations were associated and interpreted with an increasing fraction of aspherical particles, as no other information on particle complexity was available at that time.

Also in ML-CIRRUS I observed traces of optically near-spherical ice particles in the contrails (Fig. 6.21). I call these particles near-spherical because, although they show a concentric ring structure, they can be still clearly identified as ice particles. I performed a variance analysis to the dataset, similar what was done in the frozen droplet simulation experiments in Sec. 5.1.2. The analysis results showed a mean v_{az} well above the threshold value, so we can conclude that the majority of the contrail ice particles had already grown aspherical. The ages of the two contrails were estimated by Dr. T. Jurkat from DLR based on the ice particle size and number concentrations as well as NOy data. Additional information of the contrail age I got from the the Contrail Cirrus Prediction (CoCiP) model results run by Prof. U. Schumann from DLR. Based on this analysis, we can say that the first contrail was 30–45 minutes old, whereas the second contrail was younger, about 10–20 minutes old. Therefore, in both cases the contrails were old enough that we cannot anymore expect to find significant fraction of spherical particles. The ice particles had probably formed as frozen droplets but developed a roughness layer in the subsequent growth, similar what was observed in the laboratory experiments.

ML-CIRRUS flight on 11 April 2014

The last ML-CIRRUS case study in this thesis is the flight on 11 April 2014. The meteorological situation on 10 April 2014 predicted the formation of a large WCB system over the Atlantic that would head towards the coast of Scotland. The scientific team decided to sample the system, and on 11 April 2014 the HALO aircraft started towards Scotland. The WCB system (Fig. 6.22) was reached at 10:10 UTC and was first sampled at an altitude of 9.2 km while heading from the Scottish coast towards west. After the edge of the system was reached, the HALO descended to an altitude of 8.6 km and returned the same route sampling the system at the same altitude towards east. Similarly to the other cases, I verified the system as a WCB with marine origin with the help of the ECMWF trajectory analysis (Fig. 6.23 panel c).

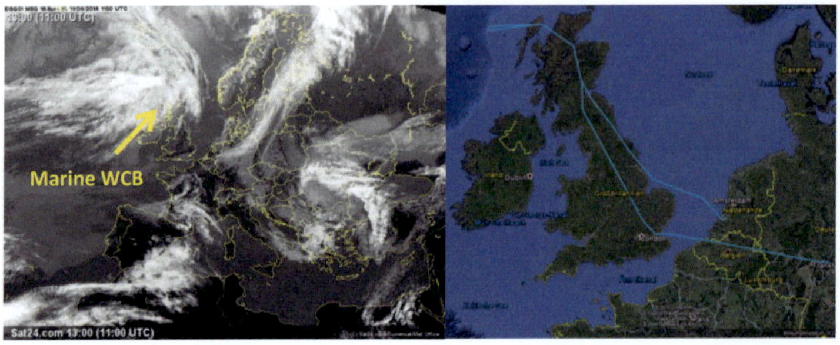

Figure 6.22.: Satellite image of the sampled marine WCB at the west coast of Scotland on 11 April 2014 (left) and the HALO flight track (right).

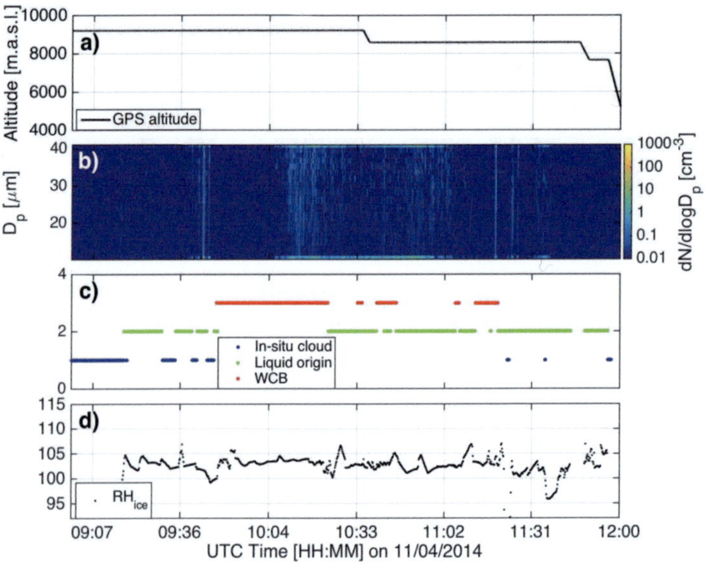

Figure 6.23.: The sampled marine WCB at the coast of Scotland on 11 April 2014. Panel (a) shows the HALO flight altitude, panel (b) the SID-3 size distribution, panel (c) the cloud origin classification based on the ECMWF analysis and panel (d) the maximum ice saturation ratio that the air parcel had experienced.

The panel b) in Fig. 6.23 shows the SID-3 size distribution from the WCB sampling. Sub-40 µm ice particles are present, although in contrary to the other case studies a clear mode of small ice particles is not seen. For this particular flight the history of the ice particle formation and growth conditions along the HALO trajectory was estimated with the ECMWF analysis. The maximum ice saturation ratio was calculated to define the formation and initial growth conditions of the ice particles (Fig. 6.23 panel d). An average maximum ice saturation of 1.03 was found for the WCB ice particles, although it has to be kept in mind that the representativeness of this number is highly uncertain. Based on the laboratory study a growth at such low saturation ratios would promote the formation of pristine particles. The presence of pristine particles was supported by a halo observation that was done while sampling the WCB (Fig. 6.24). Furthermore, the SID-3 images revealed a fraction of pristine halo producing particles, although majority of the observed ice particles were found to have a significant degree of complexity.

Figure 6.24.: A halo phenomenon photographed from the HALO aircraft (left, photograph by Dr. M. Schnaiter) and examples of SID-3 scattering patterns of halo-producing ice particles (right). The halo phenomenon can be seen on the single particle patterns as a bright spot at the edge of the image annulus (around 22°).

6.2.2. The small-scale complexity

Similar to the ACRIDICON-CHUVA campaign, I studied the small-scale complexity of the mid-latitude ice particles measured during the ML-CIRRUS flights using the SID-3 method. Figure 6.25 shows the results of this analysis. In general, the lowest fractions of complex particles is observed in the two WCB cirrus cases. The continental WCB cirrus shows lower small-scale complexity than the maritime origin WCB cirrus. The decrease in the small-scale complexity can again be explained by the aerosol effect, similar to what was observed during the ACRIDICON-CHUVA campaign. The continental WCB was influenced by the Saharan dust outbreak and the increased aerosol load could have promoted ice formation and growth at already lower supersaturations.

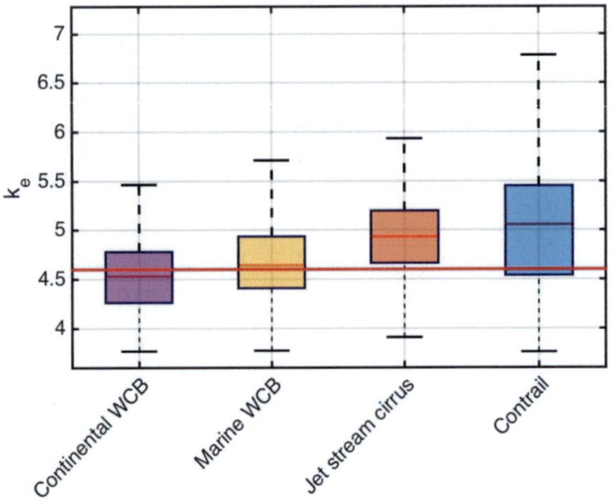

Figure 6.25.: The statistical analysis of ice particle small-scale complexity of the different case studies presented in this thesis.

As mentioned before, the maximum ice supersaturation during the ice particle growth was estimated in the marine WCB case study on 11 April 2014. The low ice saturation ratio of 1.03 would explain the relatively high fraction of pristine particles and the halo observation, although the fraction of complex particles was still higher than expected, 54%. Therefore, it is apparent that a fair amount of work is needed to understand the formation of ice particle complexity in natural cirrus clouds.

Even though over half of the ice particles were complex on 11 April 2014, we were able to observe a halo that is normally considered as a product of pristine ice particles. The formation of the halo phenomenon was studied during the ML-CIRRUS campaign in the PhD work of Linda Forster from the University of Munich. She found out using a ray tracing model that only approximately 20% of the ice particles need to be pristine to produce a visible halo. I contributed to this analysis by calculating the fraction of single ice particles that produces a visible halo spot in the SID-3 images (Fig. 6.24) and derived a fraction of 8–27%, which agrees well with the ray tracing results.

The highest degree of complexity was observed in the jet stream in-situ cirrus and in the contrails. The jet stream brought relatively clean marine air, which could promote homogeneous nucleation at high altitudes. This, as also observed during ACRIDICON-CHUVA, could promote enhanced ice particle complexity. From all the presented case studies, the highest complexity was observed in contrail cirrus, although simultaneously about 25% of the ice particles appeared pristine. This span can be explained by the simultaneous appearance of smooth quasi-spherical frozen droplets and aspherical ice particles. The frozen droplets laboratory study in Sec. 5.1.2 showed that frozen droplets exhibit the largest complexity range depending on their surface properties. Similarly, in the studied contrails we saw evidence of smooth optically near-spherical frozen droplets but also complete distorted particles were found.

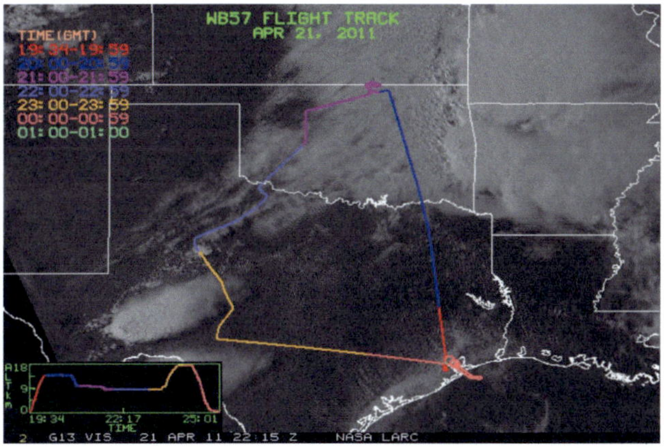

Figure 6.26.: The flightpath of the NASA WB-57 aircraft over-plotted on a satellite image. The anvil sampling was conducted between 23:00 UTC and 23:59 UTC and is marked with orange colour. The two convective systems are seen in the lower left corner of the satellite image.

6.2.3. Case of optically spherical ice in a mid-latitude thunder storm

Besides in contrails, occasionally traces of optically near-spherical particles were found in the WCB cirrus but never enough that these particles could be considered statistically significant. The updraft speeds in the WCB systems were likely not high enough that large amounts of supercooled water would have survived to the homogeneous freezing threshold, where it would have frozen simultaneously as frozen droplets. Such behaviour can only been seeing in strong convection linked to thunderstorms or tropical convective systems. Although the convection in the tropics is strong enough, frozen droplets are rarely observed in high numbers, whereas plate-like particles dominate the anvil microphysics (Stith et al., 2002; Lawson et al., 2003; Frey et al., 2011). For example, during ACRIDICON-CHUVA campaign, we had only one case, where quasi-spherical ice particles were observed from the PHIPS-HALO images.

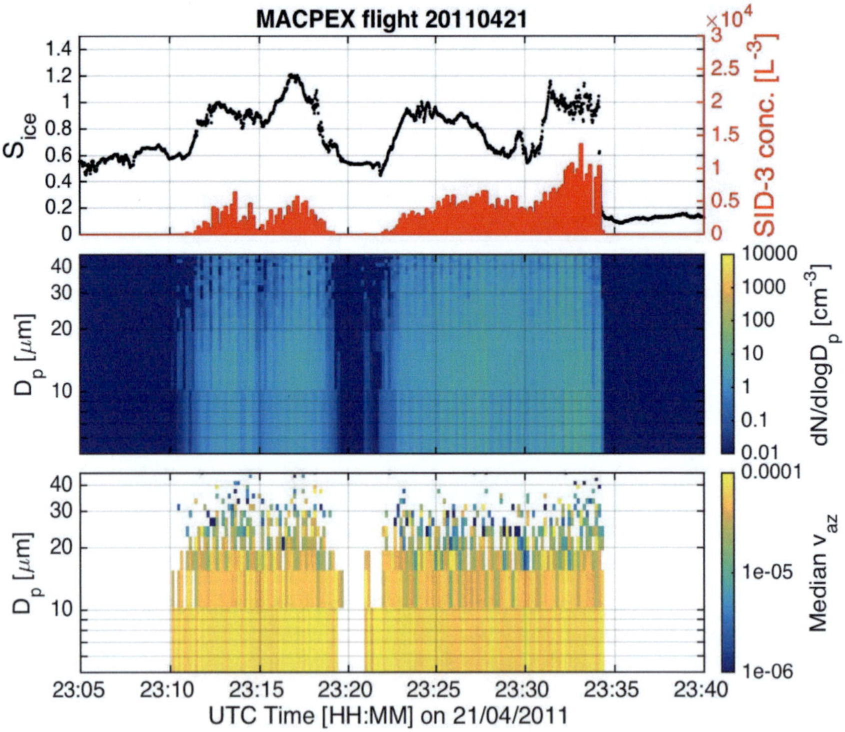

Figure 6.27.: SID-3 and water vapour measurements during anvil sampling on 21 April 2011. The upper panel shows the total concentration for particles between 5 and 45 µm and the ice saturation ratio S_{ice}. The second panel shows the particle size distribution and the third panel the size segregated variance analysis results.

The formation of frozen droplets is more common in mid-latitude thunderstorms (Gayet et al., 2012a; Stith et al., 2014). For this thesis I analysed one case, where a mid-latitude convective system was sampled over Texas on 21 April 2011 during the Mid-latitude Airborne Cirrus Properties Experiment (MACPEX) campaign with the NASA WB-57 aircraft. On that day two convective systems had developed over western Texas and northern Mexico, and the anvil outflow from these two systems extended ~100 km east towards central Texas. Fig. 6.26 shows the flight path of the WB-57 over-plotted on a satellite image. The measurements in the anvil outflows were done between 23:00-23:59 UTC (flight path marked in orange). The WB-57 flew first under the northern anvil at an altitude of 9 km sampling the lower part of the outflow. At 23:22 UTC the aircraft ascended through the anvil outflow exiting at an altitude of 13.5 km. The temperature at the lower parts of the anvil was −34°C and a temperature of −61°C was measured at the upper part of the anvil.

The anvil profile during the ascent shows an increase in the sub-40 µm ice particle concentrations from 3.5 cm^{-3} to 10 cm^{-3} (Fig. 6.27). Particularly small ice crystals in the sub-20 µm size rage are found at the top of the convective system. Although the ice particle concentrations were significantly higher than normally measured in cirrus clouds (Lawson et al., 2006; Krämer et al., 2009), they compared to what was previously measured for a mid-latitude storm (Gayet et al., 2012a; Stith et al., 2014). An inspection of the CPI images (Fig. 6.28) revealed small ice particles with a distinct signature of frozen droplets. Approximately 84% of all the CPI observed particles were classified as single frozen droplets and 2.2% as aggregates of frozen droplets. The high fraction of frozen droplets indicate that the majority of the ice particles were formed through liquid phase in the mixed-phase region of the convective cell and were transported to the anvil region in the updraft, where they remained the dominant particle type. The high fraction of frozen droplets is in agreement with previous studies of mid-latitude convective systems (Gayet et al., 2012a; Stith et al., 2014).

The water vapour measurements during the anvil sampling with the NASA Diode Laser Hygrometer (DLH; Diskin et al. (2002)) showed that most of the time the frozen droplets were found in ice sub-saturated conditions (upper panel in Fig. 6.27). Especially strong sub-saturated conditions were measured around 23:30 UTC, when S_{ice} around 0.6 was measured. The sub-saturated conditions together with the high concentration of small ice particles give evidence that these ice particles had formed almost simultaneously in a vigorous

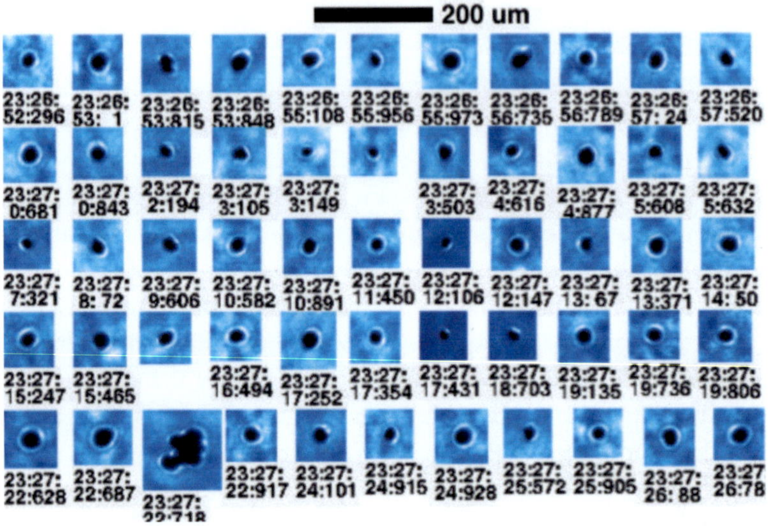

Figure 6.28.: A collection of ice particles imaged by the 3V-CPI in the convective outflow during MACPEX. Mainly single frozen droplets were detected with few aggregates of frozen droplets. Courtesy Dr. P. Lawson, SpecInc.

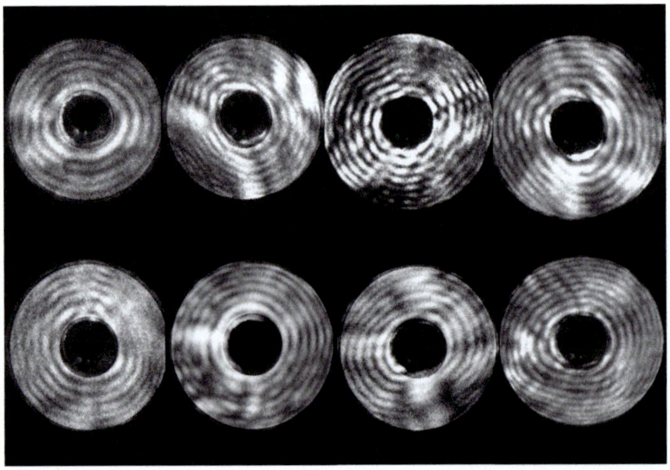

Figure 6.29.: Examples of SID-3 scattering patterns of optically near-spherical ice particles measured in the upper anvil layer of the mid-latitude thunderstorm over Texas.

updraft, and thus quickly depleted the available water vapour. This prevented further ice particle growth after freezing, and the particles remained small. The same phenomenon was also observed in the laboratory experiments: after the simulated cloud was glaciated, the S_{ice} started to quickly decrease although the chamber cooling was continued (Sec 5.1.2). The results from the chamber experiments indicate that the majority of the frozen droplets measured by the CPI during the MACPEX flight should have been in sublimating state, and we would expect to see spherical ice particles to appear. An inspection of the SID-3 diffraction patterns indeed shows signatures of spherical particles (Fig. 6.29). SID-3 images with clear ring patterns were observed in the convective outflow, similar to that, what was observed in the laboratory in case of sublimating frozen droplets. The presence of sublimating frozen droplets was also confirmed with the automated variance analysis that showed low v_{az} values for ice particles around 20 µm (blue to green colours in third panel of Fig. 6.27). The size of the smooth frozen droplets seen by SID-3 is comparable to the size of the frozen droplets seen in the CPI images, and the measured frozen droplet sizes also agree with previous observations (Baran et al., 2012b; Stith et al., 2014).

6.3. The global picture of ice crystal complexity

So far we have discussed the ice particle complexity for different case studies portraying mid-latitude and tropical cirrus systems with varying ice formation pathways. We have gotten indications that independent of the cirrus origin or location we can expect a significant degree of ice particle complexity. To form a more global picture of cirrus ice particle complexity, I have advanced the statistical analysis presented in this chapter to cover all the measurement campaigns, where SID-3 was operational. From the aforementioned ACRIDICON-CHUVA and ML-CIRRUS campaigns I have now included every measured cirrus ice crystal to this analysis.

To advance the dataset to cover also arctic clouds, I analysed two flights from the RACEPAC campaign that took place in spring 2014. In contrary to the other campaigns the ice particles in RACEPAC were found mainly in mixed-phase environment. The assumptions made of mixed-phase ice particles can have more severe consequences for the cloud radiative forcing than the cirrus ice particles (Ulrike Lohmann, personal communication 2015) and it has been shown that the radiative properties of mixed-phase clouds are highly sensitive

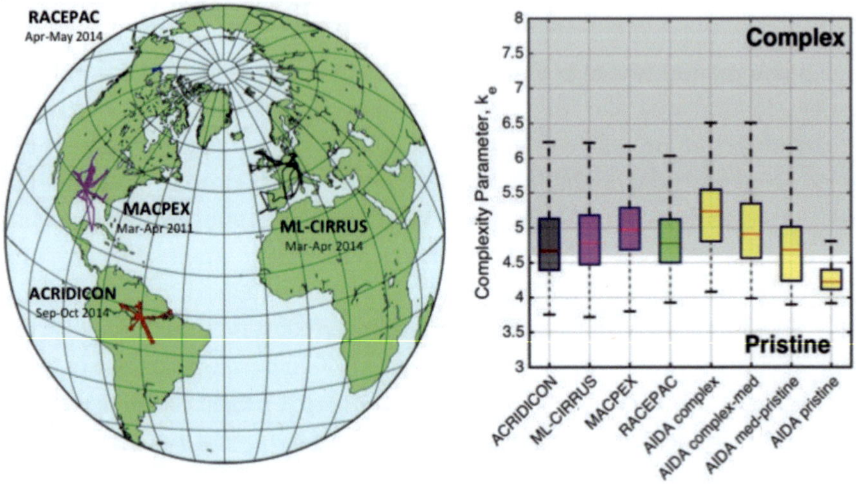

Figure 6.30.: The statistical analysis of ice particle complexity in the tropical campaign ACRIDICON-CHUVA (black), in the mid-latitude campaigns ML-CIRRUS and MACPEX (purple), in the arctic campaign RACEPAC (green), and in the AIDA simulation experiments (yellow).

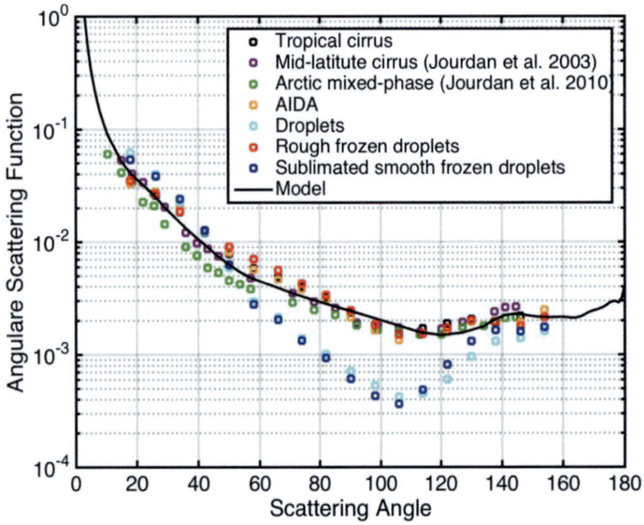

Figure 6.31.: The averaged angular scattering functions for tropical, mid-latitude and arctic ice particles as well as for AIDA simulated cirrus clouds and frozen droplets and a modelled scattering phase function.

to ice crystal morphology (Ehrlich et al., 2008), especially if small ice crystals are present.

The deployment and development of the SID-3 method to discriminate small ice particles in mixed-phase clouds during RACEPAC was part of the PhD work by Paul Vochezer and the first results are shown in his PhD thesis (Vochezer, 2015). In my PhD work I advanced the arctic mixed-phase cloud dataset by performing complexity analysis of the SID-3 scattering patterns for selected flights. In mixed-phase clouds the high particle number concentrations evidently lead to coincidence sampling in the SID-3 (as discussed in Sec. 3.2.3), which makes identifying ice particles through automated routines impossible. Only through manual selection the ice particles can be identified and analysed. In this thesis I performed a manual cross check for two RACEPAC flights and could include the first results of the complexity analysis of natural mixed-phase clouds using the SID-3 method. I analysed the flight number 2 on 30 April 2014 and flight number 6 on 8 May 2014, where a sufficient number of small ice particles were detected by the SID-3. A broader analysis of all the RACEPAC flights and other mixed-phase campaigns is an ongoing work.

Figure 6.30 shows the results of the statistical analysis of the complexity parameter, k_e, for the different field campaigns in comparison with the results from the AIDA cloud chamber. Based on the previously presented case studies, it is not a surprise that we observe a high degree of complexity in every campaign independent of the location and formation pathways of the clouds (liquid origin, in-situ, WCB or deep convection). Also, in the case of the arctic mixed-phase clouds, the ice particle complexity is comparable to the complexity found in the other campaigns, although it has to be kept in mind that the whole RACEPAC dataset is not yet analysed. Results from the cloud chamber studies indicate that the complexity of mixed-phase ice particles can be enhanced due to their rapid growth at near water saturation. Also, the ACRIDICON-CHUVA results give evidence that ice particle fromation, growth, and interaction processes in the mixed-phase cloud conditions can significantly enhance the ice particle complexity.

Albeit some small differences are seen between the complexity values of different campaigns, the median k_e value (red bar) is always above the threshold value of 4.6, which, in laboratory experiments, led to uniform angular light scattering properties. Furthermore, the complexity of the natural ice particles compare well to the laboratory experiments with moderate to complex ice particles (AIDA complex-med). Assuming only smooth (AIDA pristine) or severely complex (AIDA complex) ice particles do not seem to represent the natural ice particles. Similarly, the particle models that best fit the satellite observations always require a mixture of smooth and complex (roughened) ice particles (Cole et al., 2013; Liu et al., 2014).

The initial underlying question of this thesis has been, how does the particle complexity, here in the case of ice particles, influence the angular light scattering properties, i.e. the asymmetry parameter. In the case of ice clouds, the asymmetry parameter is one of the key features that determines the cloud short-wave radiative properties and consquently the cloud radiative forcing. Thus, it is only appropriate that the last illustration of this thesis summarises the observed ice particle angular scattering functions from the tropics to the arctic and in the simulated ice clouds (Fig. 6.31). For the tropical and laboratory cases the averaged angular scattering functions were measured with the PHIPS-HALO instrument and the particle complexity was simultaneously defined with the SID-3 method. As for the mid-latitude and arctic cases PHIPS-HALO scattering data was not available, I relied on the PN measurements of the ice particle angular scattering properties. As a representative mid-latitude ice cloud angular scattering function I used the PN measurements from CIRRUS 98 campaign over France (Jourdan et al., 2003) and for an arctic ice cloud angular scattering function the PN measurements from the top of an arctic nimbostratus cloud during the ASTAR 2004 experiment (Jourdan et al., 2010). For comparison purposes, I also show the angular scattering function of a liquid cloud measured during the ACRIDICON-CHUVA campaign with the PHIPS-HALO instrument.

Again, we see a little variation between the angular scattering functions measured for the different locations, with the exception of the laboratory angular scattering function of sublimating frozen droplets. The different angular scattering functions tend to cluster on one functional dependency, although small variations are seen in the forward scattering angular range below 80°. This, however, could also be of instrumental origin as the largest variation is seen between the scattering functions measured by the two different instruments. It can be concluded that Fig. 6.30 supports the hypothesis that ice particle complexity is a common feature and this is conveyed into a globally uniform ice particle angular light scattering function with enhanced scattering into the backward hemisphere. Therefore, the global ice cloud scattering function might largely be dominated by the degree of ice particle complexity. Ice particle shape does not seem to influence the scattering properties significantly, after a certain threshold complexity is achieved. This was clearly demonstrated by the simulation experiments with frozen droplets: the complex frozen droplets showed an angular scattering function that resembled that of any other of complex ice particles but if the complexity was removed from the particle surface in sublimation, the scattering function of the smooth ice spheres was again governed by their shape, i.e. we had a similar scattering behaviour than that of liquid droplets (spheres).

The measured angular scattering functions cover only a limited angular range and, therefore, to define an asymmetry parameter we need to extrapolate the measurements to cover the full angular range. This was done in cooperation with Dr. O. Jourdan from University Blaise Pascal. We used the scattering database of Prof. P. Yang to form a particle model using a combination of spherical particles (effective diameter from 1 to 250 µm) and deeply rough hexagonal plates with an aspect ratio of 0.5 (maximum dimensions from 20 to 900 µm). The advantage of using a particle model is that we can reduce the uncertainty of the retrieved asymmetry parameter compared to using an automated algorithm. The retrieved asymmetry parameter was 0.796 ± 0.02, which is comparable to previous measurements in ice clouds (e.g. Gayet et al., 2004; Febvre et al., 2009; Cole et al., 2014), but lower compared to the asymmetry parameters of pristine ice particle habits in Prof. Yang's database (from 0.8 to 0.95, depending on the habit and size).

Although the change in the asymmetry parameter might be small, this can have severe consequences for the cloud radiative forcing. In a study of Yi et al. (2013) the influence of ice particle complexity to the global cloud radiative effect was investigated and the authors found out that just by changing the surface properties of the ice particles in a climate model created a cooling effect of $1-2\,\mathrm{W\,m^{-2}}$, which is comparable to the CO_2 forcing, but opposite in sign. The authors used completely smooth pristine ice crystals as the reference case, so the above forcing range is just an upper estimate for the climate impact of ice particle complexity. However, as it is demonstrated in this thesis, a low asymmetry parameter might be a common feature in natural ice clouds, which may have severe consequences for our climate. The possible implications should be addressed in future studies.

7. Summary and Outlook

The main objective of this thesis was to experimentally study the angular light scattering properties of three complex atmospheric particle systems: ice crystals, mineral dust aerosol and viscous secondary organic aerosol. I presented three types of light scattering measurements covering different angular ranges: the near-backscattering (176° or 178°) depolarisation measurements, the spatially-resolved scattering measurements at the near-forward angular range between 6° and 24° and the scattering function measurements covering the angular range from 18° to 170°. The measurements were conducted in the cloud chambers AIDA and CERN CLOUD as well as during field studies in the tropics, mid-latitude and in the arctic.

The results of this thesis will contribute to the understanding of the short-wave scattering properties of atmospheric particles and will improve the knowledge on the origin and microphysical details of smalls and large-scale ice particle complexity. Furthermore, the findings presented here will provide several suggestions for future research, as will be discussed at the end of this chapter. The main scientific contributions are summarised below as listed in the introduction:

Measurements of near-backscattering depolarisation ratio. In the context of this PhD project I built and characterised an in-situ scattering and depolarisation instrument to be operated at the AIDA and CERN CLOUD chambers. I was able to demonstrate that near-backscattering depolarisation measurements are highly sensitive to even small deviations from the spherical shape already at particle sizes slightly above 100 nm. These measurements were conducted with α-pinene SOA particles in cloud chamber experiments where I deployed depolarisation measurements to detect particle anisotropy or shape deformation caused by the viscous phase state of these particles. Furthermore, the phase transformation to low-viscous states could be clearly detected by a decrease of the linear and circular depolarisation ratios.

The second studied aerosol system was dust particles dispersed in a cloud chamber. I used a novel dispersion technique to generate narrow size distributions of dust particles and used samples from several sources covering desert dusts from Asia and Africa and soil samples from South-America, Europe and North-Africa. The size range that was covered by these experiments ranged from 0.33 µm to 3 µm in particle volume median diameter, complementing previously published size-segregated laboratory measurements of dust aerosol. The main finding was that the dust depolarisation properties are highly dependent on the particle size. Fine-mode dust has depolarisation ratios below 0.3, whereas coarse-mode particles show higher but stable depolarisation ratios that is on average around 0.27. Little difference was seen in the depolarisation properties between the different dust samples, although their source regions and mineralogical composition was assumed to be variable.

Also, in the case of ice particles their depolarisation properties were largely determined by the particle size. Changes in the ice particle complexity did not seem the affect the depolarisation properties significantly. Therefore, it is indicated that although backscattering depolarisation measurements can be considered as highly sensitive to deviations from the spherical shape, they are largely insensitive to changes in particle complexity. Thus, a better measure is needed for the detection of complex particles.

Angular light scattering of complex ice particles. For larger, super-micron particles it was shown that the spatial distribution of the forward scattered light provided information of the particle morphology and offered a method to quantify the degree of particle complexity through an optical complexity parameter k_e. The method was applied in laboratory experiments to investigate the origin of ice particle small-scale complexity in the case of hexagonal and near-spherical ice particles. It was found that the ice particle small-scale complexity is driven by the ice particle growth conditions, i.e. the amount of available condensable water vapour. After a certain threshold complexity was achieved, the information of the underlying particle shape was lost and the forward scattered light intensity was uniformly distributed in space.

Although the complexity parameter, k_e, still exhibited changes with increasing degree of complexity, the angular scattering function was observed to be insensitive to further increases in the complexity after the threshold complexity was achieved. The angular scattering function was observed to be featureless with an enhanced degree of backscattering (i.e. a low asymmetry parameter). Also, no significant difference in the angular scattering functions was observed between complex hexagonal columns and complex near-spherical ice particles, indicating that complexity largely determined the ice particle scattering properties and not the particle shape. The shape dependent features in the angular scattering function were visible only if the complexity was reduced to a minimum, as shown in the case of sublimating frozen droplets.

Field observations of ice particle complexity. The knowledge gained from the laboratory studies was implemented in field studies covering ice clouds from the tropics, to mid-latitudes and reaching to the arctic. In this thesis, I presented the first comprehensive global measurements of the ice particle small-scale complexity using the SID-3 method. I also demonstrated the possibilities of the SID-3/PHIPS-HALO -instrument package to simultaneously investigate the ice particle complexity (both small-scale and large-scale) and the cloud averaged scattering properties. It was found that a similar and high degree of particle small-scale complexity was present in every measurement campaign independent of the measurement location or cloud type. The median complexity parameter, k_e, was observed to be always above the laboratory defined threshold for complex particles. This high degree of particle complexity led to a stable and almost universal angular scattering function that showed only little variations over the globe. Based on the measurement a global asymmetry parameter of 0.796 ± 0.02 was estimated.

7.1. Climate implications and outlook

There are many factors influencing the uncertainty of ice clouds in climate predictions. Particle complexity and its influence on the asymmetry parameter is one of them but other uncertainties should not be neglected, like the appearance of small ice particles and the distribution of ice sizes and habits as a function of distance from the cloud top. Until recently, the general opinion was that the ice particle asymmetry parameter is highly variable depending on the particle size, shape and wavelength. In this thesis, it was indicated that the variability in the asymmetry parameter due to changes in particle shape or size is probably smaller than previously thought. To estimate the possible impact on climate predictions, it would be recommended to incorporate the experimentally derived asymmetry parameter into climate models. For this purpose, a co-operation is already started with Prof. U. Lohmann's group at ETH Zürich. Moreover, the presented measurements give information of the asymmetry parameter only at one wavelength band and, thus, to improve the parametrisation of the asymmetry parameter in climate models, the presented measurements should be extended to other spectral bands.

This thesis showed that state-of-the-art in-situ measurements can reveal new information of the ice particle microphysics. Although it is highly important to continue with in-situ measurements and to develop new optical methods, it should be kept in mind that in-situ measurements are only snapshots of the current situation. The sampling volume of a single could probe is hardly statistically significantly; and to make statistically relevant observations we need hundreds or thousands of flight hours. Remote sensing methods provide the only method to get globally covered information of the atmospheric composition on a daily basis. Therefore, the focus should be in combining the in-situ and remote sensing observations to answer questions like: is particle complexity significant in all ice clouds over the globe?, what is the size distribution of small ice crystals?, what is the optical depth of the clouds?, etc.

In order to study the particle morphology, like complexity, we need to have a measure that is sensitive to small changes in the particle structure. The angular scattering function or the backscattering depolarisation measurements seemed to be insensitive to such changes after a certain degree of complexity was reached. However, satellites with polarisation detection have shown that polarisation measurements in the angular range from $130° - 170°$ can be sensitive to the complexity of the particles. To validate these observations and to advance the detection of particle complexity we need in-situ polarisation measurements in the given angular range, e.g. polarisation resolved angular scattering measurements.

The strength of this work was that I was able to combine laboratory experiments with atmospheric measurements. I illustrated that in order to better interpret atmospheric results, we need to take a step back and investigate the phenomena under controlled laboratory conditions. For example, the laboratory results of dust depolarisation ratios will be of use for the lidar community to develop and test lidar retrieval methods that tackle the backward problem of atmospheric aerosol particles. Especially, the results can encourage scientists to develop new algorithms to distinguish between fine- and coarse-mode dust particles. Such algorithms could improve the understanding of long range dust transport or the detection of fine-mode dust particles in forest-fire events.

I hope that after reading this work it is obvious that laboratory experiments on the optical properties of atmospheric particles are vital for developing optical particle models and increasing our understanding of atmospheric particles.

Dedicated optical experiments are needed in the future to cover the full spectrum of atmospheric light scattering, and not only in the visible spectral range but also in the infrared region. Also, the exchange of knowledge between the experimentalist and the modellers is crucial for developing the field further. At the end of this thesis, we are not left without open questions but we are on the right path to better understand the optical properties and the microphysical nature of several atmospheric particles.

Appendix

A. Laboratory characterisation of the SIMONE-Junior instrument

A.1. Laser stability

The stability of the SIMONE-Junior OBIS-laser was tested by operating the laser at the maximum power of 20 mW over several time periods. During these period the laser power was constantly measured with a PowerMax laser power sensor (PowerMax-RS, Coherent Inc.). Figure A.1 shows the laser power measurements over a time period of one hour. A median laser power of 22.11 mW was measured and the standard deviation was 0.096 mW, which is 0.4% of the measured laser power. As the depolarisation measurements are normally averaged over a time period, can it be assumed that the small variability of the laser power is averaged out and does not affect the time integrated depolarisation measurements.

A.2. Laser divergence

The OBIS laser has a divergence of 0.07°, which means that after traveling across the AIDA or CERN CLOUD chamber the one millimetre laser beam would have increased its diameter by almost a factor of 2. To get a collimated beam over the sampling volume, I used a beam expander that increases the beam size by a factor of ten and, therefore, reduces the laser divergence by the same factor. I determined the diameter of the expanded laser beam by measuring the laser profile at two distances by using a laser profile camera (LaserCam-HR, Coherent Inc.). The laser diameter increased its size by 0.2 mm in a distance of 2844 mm. With these numbers we can determine a laser divergence of 0.002°. This means that the laser beam will change its diameter only by 0.28 mm after traveling across the chamber, which is around 3% change in the beam size.

A.3. Acceptance angle of the telescopes

The scattered light from the centre of the chamber is collected with two identical telescopes. The telescope works as a refractive telescope with an ocular lens and with an objective lens. The focal length of the ocular lens is 50 mm and the objective lens 200 mm, which creates a 4-fold magnification. At the focal point of these two lenses is a pinhole with a 0.3 mm diameter. The radius of the pinhole ($r_{pinhole}$), together with the focal length of the objective lens ($f_{objective}$) determine the acceptance angle of the telescope through

$$\alpha = arctan\left(\frac{r_{pinhole}}{f_{objective}}\right). \tag{A.1}$$

With the given values, we get a accepting angle of $\alpha = 0.043°$ for both the forward and backward telescopes.

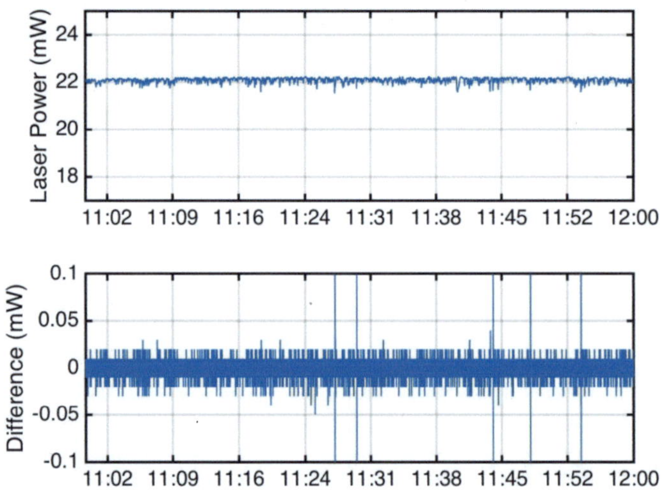

Figure A.1.: Laser stability measurements.

A.4. Detection angle and detection volume

Since the SIMONE-Junior uses continuous wave laser, the backward measurements cannot be conducted at the exact backward direction (180°) but at small offset from this angle. The offset from the exact backward and forward angles (β) is determined by the geometry of the setup and is the same for both backward and forward directions, as the forward and backward telescopes are aligned so that they are facing each other. Here, I determine the detection angle of the SIMONE-Junior instrument geometrical using the chamber dimensions. The distance between the laser entry and the backward telescope, a, is the same for both AIDA and CERN CLOUD chambers, 10 cm. The distance between the laser entry and the chamber center, L, or the detection telescope and the chamber center, D, is defined by the chamber radius, the wall thickness, the thickness of the isolating housing and the length of the isolating tube, where the instrument is mounted. For the AIDA chamber $L = D = 322$ cm, and for the CERN CLOUD chamber $L = D = 150$ cm. The measurement angle can be determined through

$$\beta = arcsin\left(\frac{a}{D}\right). \tag{A.2}$$

For the AIDA chamber the difference from the exact backward direction is 1.78° and for the CERN CLOUD chamber 3.8°.

Also the detection volume can be defined through the chamber geometry (Fig. A.3). The laser beam has a diameter of 8 mm in the center of the AIDA chamber. The length of the overlap region of the laser and the detection aperture, c, was approximated by illuminating the detection telescope from behind and by following the overlap with a scattering target in the center of the chamber. An estimated length for c covers 600 mm. Now the detection volume can be calculated through

$$V = \frac{\pi}{4}d^2c = 30cm^3. \tag{A.3}$$

In CERN CLOUD chamber I could not estimate the length c, as the chamber cannot be entered. However, we can assume that the detection volume in the CERN CLOUD chamber is comparable to the AIDA chamber. In our case it is not important to know exactly, how many particles we have in the detection volume. Important is to know that already at relatively low concentrations around few tens of particles per litre, we have on average at least one particle in the detection volume. This means that at typical chamber concentrations of 1-1000 cm^{-3} we have always multiple particle simultaneously in the detection volume, which means that we have enough statistics to assume random orientation of the particles.

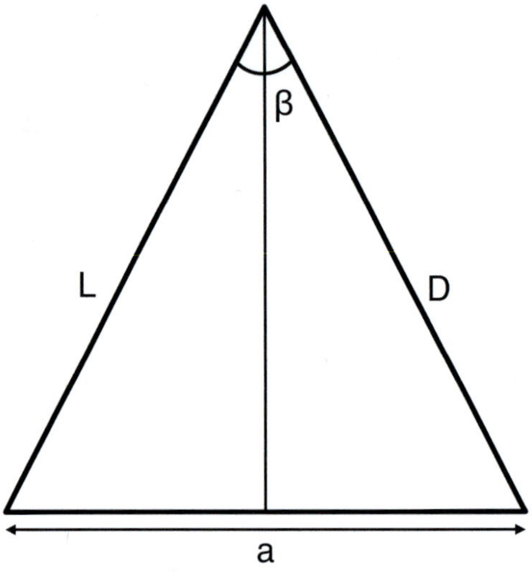

Figure A.2.: The geometry of the SIMONE-Junior setup in an cloud chamber. The same geometry is valid for the forward and backward detection.

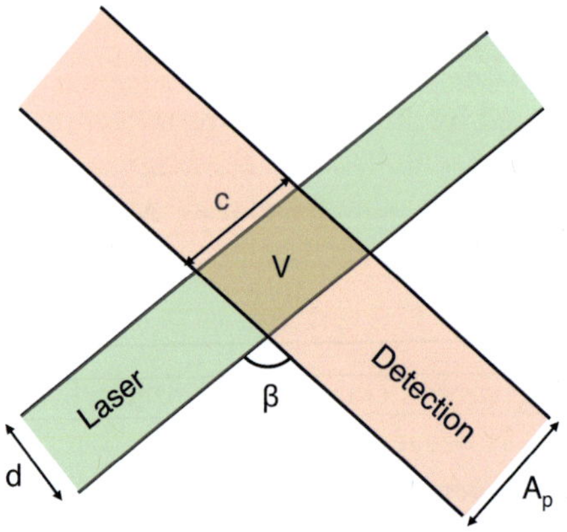

Figure A.3.: The detection volume that is defined by the laser beam and by the detection aperture.

A.5. Calibration of the neutral density filters and the laser line filters

SIMONE-Junior has a set of neutral density filters with optical thickness of 1 that are installed in front of the backward and the forward photon multipliers in a filter wheel (forward) or in a filter flip (backward). An automated program was written that introduces a neutral density filter in front of the photon multiplier, when an intensity $<10^6$ counts is measured. If the signal is found below 10^5 counts, the filter is removed. The highest intensity is measured in the forward direction, so an additional neutral density filter with optical thickness of 2 was added that can be replace the neutral density filter with optical thickness 1, if the intensity exceeds 10^{12} counts.

I determined the actual transmission of the filters in a laboratory calibration. The calibration step consisted of an Ulbricht-sphere, a neutral density filter with optical thickness of 7, the SIMONE-laser and a photon multiplier. The SIMONE-Junior 552 nm laser was directed to the Ulbricht-sphere and the intensity of the laser was decreased with an neutral density filter before it was measured with a photon multiplier. I first measured the laser intensity with this setup and then added one by one the neutral density filters before the photon multiplier. The reduction in the laser power was measured and this number gave the actual transmission of the neutral density filters (Table A.1).

Additionally, the photon multipliers have laser line filters that also affect the transmission. I measured transmissions around 0.4 for the laser line filters with some small differences between the individual filters. However, it is not necessary to correct for these differences, because the photon multiplier correction factor includes this correction.

Backward filter, optical thickness 1	0.0974
Forward filter, optical thickness 1	0.1255
Forward filter, optical thickness 2	0.0128

Table A.1.: The measured values for the transmission of the neutral density filters.

B. Derivation of the linear and circular depolarisation ratios

In this appendix I derive the linear and circular depolarisation ratio measured by the SIMONE instruments using the scattering matrix formalism.

B.1. Linear depolarisation ratio

SIMONE uses linearly polarised light with polarisation vector aligned parallel, $\vec{I_i} = [1, 1, 0, 0]$, or perpendicular, $\vec{I_i} = [1, -1, 0, 0]$, to the scattering plane. Now, if we consider an ensemble of particles with random orientation, the scattered Stokes vector can be written as

$$\begin{pmatrix} I_s \\ Q_s \\ U_s \\ V_s \end{pmatrix} = \frac{1}{k^2 r^2} \begin{pmatrix} S_{11}(\Theta) & S_{12}(\Theta) & 0 & 0 \\ S_{12}(\Theta) & S_{22}(\Theta) & 0 & 0 \\ 0 & 0 & S_{33}(\Theta) & S_{34}(\Theta) \\ 0 & 0 & -S_{34}(\Theta) & S_{44}(\Theta) \end{pmatrix} \begin{pmatrix} I_i \\ Q_i \\ U_i \\ V_i \end{pmatrix}. \tag{B.1}$$

In the case of parallel incident polarisation, the scattered Stokes vector gets the form

$$\begin{pmatrix} S_{11} + S_{12} \\ S_{12} + S_{22} \\ 0 \\ 0 \end{pmatrix} \tag{B.2}$$

and in the case of perpendicular incident polarisation

$$\begin{pmatrix} S_{11} - S_{12} \\ S_{12} - S_{22} \\ 0 \\ 0 \end{pmatrix} \tag{B.3}$$

The scattered light is passed through a Wollaston prism that divides the scattered light to its parallel and perpendicular polarisation components. The Müller matrix for the two transmission are

$$\frac{1}{2}\begin{pmatrix} 1 & 1 & 0 & 0 \\ 1 & 1 & 0 & 0 \\ 0 & 0 & 0 & 0 \\ 0 & 0 & 0 & 0 \end{pmatrix} \tag{B.4}$$

$$\frac{1}{2}\begin{pmatrix} 1 & -1 & 0 & 0 \\ -1 & 1 & 0 & 0 \\ 0 & 0 & 0 & 0 \\ 0 & 0 & 0 & 0 \end{pmatrix}, \tag{B.5}$$

respectively. By applying these Müller matrices to the scattered light, we get the following intensity components in the case of parallel incident polarisation

$$I_{\parallel} = \frac{1}{2}\begin{pmatrix} S_{11} + 2S_{12} + S_{22} \\ S_{11} + 2S_{12} + S_{22} \\ 0 \\ 0 \end{pmatrix} \tag{B.6}$$

$$I_{\perp} = \frac{1}{2}\begin{pmatrix} S_{11} - S_{22} \\ S_{22} - S_{11} \\ 0 \\ 0 \end{pmatrix}. \tag{B.7}$$

and in the case of perpendicular incident polarisation

$$I_{\parallel} = \frac{1}{2}\begin{pmatrix} S_{11} - S_{22} \\ S_{11} - S_{22} \\ 0 \\ 0 \end{pmatrix} \tag{B.8}$$

$$I_{\perp} = \frac{1}{2} \begin{pmatrix} S_{11} - 2S_{12} + S_{22} \\ -S_{11} + 2S_{12} - S_{22} \\ 0 \\ 0 \end{pmatrix} . \quad \text{(B.9)}$$

The linear depolarisation ratio in the case of parallel incident polarisation is now

$$\delta_{L,h} = \frac{I_{\perp}}{I_{\parallel}} = \frac{S_{11} - S_{22}}{S_{11} + 2S_{12} + S_{22}} \tag{B.10}$$

and in the case of perpendicular incident polarisation

$$\delta_{L,v} = \frac{I_{\perp}}{I_{\parallel}} = \frac{S_{11} - S_{22}}{S_{11} - 2S_{12} + S_{22}}. \tag{B.11}$$

At exact backward direction (180°) the matrix element $S_{12} = 0$ and the linear depolarisation ratio will simplify as

$$\delta_{L} = \frac{S_{11} - S_{22}}{S_{11} + S_{22}}. \tag{B.12}$$

In the case of spherical particles (e.g. water droplets) applies $S_{11} = S_{22}$, which leads to zero depolarisation independent of the measurement angle.

B.2. Circular depolarisation ratio

In order to measure circular depolarisation ratio, we need to be first able to produce circularly polarised light. This is done by introducing a quarter wave ($\lambda/4$) plate with fast axis set to 45° angle with respect to the laser polarisation. The Müller matrix of this quarter wave plate is

$$\begin{pmatrix} 1 & 0 & 0 & 0 \\ 0 & 0 & 0 & -1 \\ 0 & 0 & 1 & 0 \\ 0 & 1 & 0 & 0 \end{pmatrix} . \tag{B.13}$$

Now we get right-handed circularly polarised light, if the laser is parallel polarised

$$\begin{pmatrix} 1 \\ 1 \\ 0 \\ 0 \end{pmatrix} = \begin{pmatrix} 1 & 0 & 0 & 0 \\ 0 & 0 & 0 & -1 \\ 0 & 0 & 1 & 0 \\ 0 & 1 & 0 & 0 \end{pmatrix} \begin{pmatrix} 1 \\ 0 \\ 0 \\ 1 \end{pmatrix} \tag{B.14}$$

and left-handed circularly polarised light, if the laser is perpendicularly polarised

$$\begin{pmatrix} 1 \\ -1 \\ 0 \\ 0 \end{pmatrix} = \begin{pmatrix} 1 & 0 & 0 & 0 \\ 0 & 0 & 0 & -1 \\ 0 & 0 & 1 & 0 \\ 0 & 1 & 0 & 0 \end{pmatrix} \begin{pmatrix} 1 \\ 0 \\ 0 \\ -1 \end{pmatrix}. \tag{B.15}$$

The Stokes vector of the scattered light in the case of right-handed circularly polarised light gets the form

$$\begin{pmatrix} S_{11} \\ S_{12} \\ S_{34} \\ S_{44} \end{pmatrix} = \frac{1}{k^2 r^2} \begin{pmatrix} S_{11}(\Theta) & S_{12}(\Theta) & 0 & 0 \\ S_{12}(\Theta) & S_{22}(\Theta) & 0 & 0 \\ 0 & 0 & S_{33}(\Theta) & S_{34}(\Theta) \\ 0 & 0 & -S_{34}(\Theta) & S_{44}(\Theta) \end{pmatrix} \begin{pmatrix} 1 \\ 0 \\ 0 \\ 1 \end{pmatrix} \tag{B.16}$$

and in the case of left-handed circularly polarised light

$$\begin{pmatrix} S_{11} \\ S_{12} \\ S_{34} \\ -S_{44} \end{pmatrix} = \frac{1}{k^2 r^2} \begin{pmatrix} S_{11}(\Theta) & S_{12}(\Theta) & 0 & 0 \\ S_{12}(\Theta) & S_{22}(\Theta) & 0 & 0 \\ 0 & 0 & S_{33}(\Theta) & S_{34}(\Theta) \\ 0 & 0 & -S_{34}(\Theta) & S_{44}(\Theta) \end{pmatrix} \begin{pmatrix} 1 \\ 0 \\ 0 \\ -1 \end{pmatrix}. \tag{B.17}$$

Before we can analyse the scattered light, we have to change the polarisation state back to linear polarisation. We do this by introducing a quarter wave plate before the Wolaston prism. Now using combining equations B.13, B.16 and B.17 we get the Stokes vector for the scattered light before the analysis with the Wolaston prism

$$\begin{pmatrix} S_{11} \\ -S_{44} \\ S_{34} \\ S_{12} \end{pmatrix} \qquad \text{(B.18)}$$

$$\begin{pmatrix} S_{11} \\ S_{44} \\ S_{34} \\ S_{12} \end{pmatrix}, \qquad \text{(B.19)}$$

for right-handed circularly and for left-handed circularly polarised light, respectively. The circular depolarisation ratio is defined as the perpendicular polarisation component divided by the parallel polarisation component after analysis with the Wollaston prism, like in the case of linear depolarisation ratio. By combining equations B.4, B.5, B.18 and B.19, we get the circular depolarisation ratio as

$$\delta_C = \frac{I_\perp}{I_\parallel} = \frac{S_{11} + S_{44}}{S_{11} - S_{44}}. \qquad \text{(B.20)}$$

List of Figures

List of Tables

Bibliography

Abdelmonem, A., Järvinen, E., Duft, D., Hirst, E., Vogt, S., Leisner, T., Schnaiter, M., 2016. Phips-halo: The airborne particle habit imaging and polar scattering probe. part i: Design and operation. Atmos. Meas. Tech. Discuss. 2016, 1–21.

Abdelmonem, A., Schnaiter, M., Amsler, P., Hesse, E., Meyer, J., Leisner, T., 2011. First correlated measurements of the shape and light scattering properties of cloud particles using the new particle habit imaging and polar scattering (phips) probe. Atmos. Meas. Tech. 4 (10), 2125–2142.

Adler, G., Koop, T., Haspel, C., Taraniuk, I., Moise, T., Koren, I., Heiblum, R. H., Rudich, Y., 2013. Formation of highly porous aerosol particles by atmospheric freeze-drying in ice clouds. Proc. Nat. Acad. Sci. USA 110 (51), 20414–20419.

Aiken, A. C., DeCarlo, P. F., Jimenez, J. L., 2007. Elemental analysis of organic species with electron ionization high-resolution mass spectrometry. Anal. Chem. 79 (21), 8350–8358.

Aiken, A. C., Decarlo, P. F., Kroll, J. H., Worsnop, D. R., Huffman, J. A., Docherty, K. S., Ulbrich, I. M., Mohr, C., Kimmel, J. R., Sueper, D., et al., 2008. O/c and om/oc ratios of primary, secondary, and ambient organic aerosols with high-resolution time-of-flight aerosol mass spectrometry. Environ. Sci. Technol. 42 (12), 4478–4485.

Almeida, J., Schobesberger, S., Kürten, A., Ortega, I. K., Kupiainen-Määttä, O., Praplan, A. P., Adamov, A., Amorim, A., Bianchi, F., Breitenlechner, M., et al., 2013. Molecular understanding of sulphuric acid-amine particle nucleation in the atmosphere. Nature 502 (7471), 359–363.

Amiridis, V., Wandinger, U., Marinou, E., Giannakaki, E., Tsekeri, A., Basart, S., Kazadzis, S., Gkikas, A., Taylor, M., Baldasano, J., Ansmann, A., 2013. Optimizing calipso saharan dust retrievals. Atmos. Chem. Phys 13 (23), 12089–12106.

Andreae, M. O., Merlet, P., 2001. Emission of trace gases and aerosols from biomass burning. Global Biogeochem. Cycles 15 (4), 955–966.

Ansmann, A., Bösenberg, J., Chaikovsky, A., Comerón, A., Eckhardt, S., Eixmann, R., Freudenthaler, V., Ginoux, P., Komguem, L., Linné, H., et al., 2003. Long-range transport of saharan dust to northern europe: The 11–16 october 2001 outbreak observed with earlinet. J. Geophys. Res. Atmos. 108 (D24).

Ansmann, A., Mattis, I., Müller, D., Wandinger, U., Radlach, M., Althausen, D., Damoah, R., 2005. Ice formation in saharan dust over central europe observed with temperature/humidity/aerosol raman lidar. J. Geophys. Res. Atmos. 110 (D18).

Ansmann, A., Petzold, A., Kandler, K., Tegen, I., Wendisch, M., Mueller, D., Weinzierl, B., Mueller, T., Heintzenberg, J., 2011b. Saharan mineral dust experiments samum–1 and samum–2: what have we learned? Tellus B. 63 (4), 403–429.

Ansmann, A., Seifert, P., Tesche, M., Wandinger, U., 2012. Profiling of fine and coarse particle mass: case studies of saharan dust and eyjafjallajökull/grimsvötn volcanic plumes. Atmos. Chem. Phys 12 (20), 9399–9415.

Ansmann, A., Tesche, M., Seifert, P., Groß, S., Freudenthaler, V., Apituley, A., Wilson, K., Serikov, I., Linné, H., Heinold, B., et al., 2011. Ash and fine-mode particle mass profiles from earlinet-aeronet observations over central europe after the eruptions of the eyjafjallajökull volcano in 2010. J. Geophys. Res. Atmos. 116 (D20).

Bailey, M. P., Hallett, J., 2009. A comprehensive habit diagram for atmospheric ice crystals: Confirmation from the laboratory, airs ii, and other field studies. J. Atmos. Sci. 66 (9), 2888–2899.

Baran, A. J., 2009. A review of the light scattering properties of cirrus. J. Quant. Spectrosc. Radiat. Transfer 110 (14), 1239–1260.

Baran, A. J., 2012b. From the single-scattering properties of ice crystals to climate prediction: A way forward. Atmos. Res. 112, 45–69.

Baran, A. J., Connolly, P. J., Heymsfield, A. J., Bansemer, A., 2011. Using in situ estimates of ice water content, volume extinction coefficient, and the total solar optical depth obtained during the tropical active campaign to test an ensemble model of cirrus ice crystals. Q. J. Roy. Meteor. Soc. 137 (654), 199–218.

Baran, A. J., Furtado, K., Labonnote, L.-C., Havemann, S., Thelen, J.-C., Marenco, F., 2015. On the relationship between the scattering phase function of cirrus and the atmospheric state. Atmos. Chem. Phys. 15 (2), 1105–1127.

Baran, A. J., Gayet, J.-F., Shcherbakov, V., 2012b. On the interpretation of an unusual in-situ measured ice crystal scattering phase function. Atmos. Chem. Phys. 12 (19), 9355–9364.

Bateman, A. P., Belassein, H., Martin, S. T., 2014. Impactor apparatus for the study of particle rebound: Relative humidity and capillary forces. Aerosol Sci. Tech. 48 (1), 42–52.

Baum, B. A., Kratz, D. P., Yang, P., Ou, S., Hu, Y., Soulen, P. F., Tsay, S.-C., 2000. Remote sensing of cloud properties using modis airborne simulator imagery during success: 1. data and models. J. Geophys. Res. Atmos. 105 (D9), 11767–11780.

Baum, B. A., Yang, P., Heymsfield, A. J., Schmitt, C. G., Xie, Y., Bansemer, A., Hu, Y.-X., Zhang, Z., 2011. Improvements in shortwave bulk scattering and absorption models for the remote sensing of ice clouds. J. Appl. Meteor. Climatol. 50 (5), 1037–1056.

Baum, B. A., Yang, P., Hu, Y.-X., Feng, Q., 2010. The impact of ice particle roughness on the scattering phase matrix. J. Quant. Spectrosc. Radiat. Transfer 111 (17), 2534–2549.

Baumgardner, D., Newton, R., Krämer, M., Meyer, J., Beyer, A., Wendisch, M., Vochezer, P., 2014. The cloud particle spectrometer with polarization detection (cpspd): A next generation open-path cloud probe for distinguishing liquid cloud droplets from ice crystals. Atmos. Res. 142, 2–14.

Berkemeier, T., Shiraiwa, M., Pöschl, U., Koop, T., 2014. Competition between water uptake and ice nucleation by glassy organic aerosol particles. Atmos. Chem. Phys. 14 (22), 12513–12531.

Bingemer, H., Klein, H., Ebert, M., Haunold, W., Bundke, U., Herrmann, T., Kandler, K., Müller-Ebert, D., Weinbruch, S., Judt, A., et al., 2012. Atmospheric ice nuclei in the eyjafjallajökull volcanic ash plume. Atmos. Chem. Phys. 12 (2), 857–867.

Bohren, C. F., Huffman, D. R., 2008. Absorption and scattering of light by small particles. John Wiley & Sons.

Boucher, O., Randall, A. D., Bretherton, P., Feingold, C., Forster, G., Kerminen, P., Kondo, V., Liao, Y., Lohmann, H., Rasch, U., et al., 2013. Clouds and aerosols in climate change 2013. In: The Physical Science Basis. Contribution of Working Group I to the Fifth Assessment Report of the Intergovernmental Panel on Climate Change. Cambridge University Press Cambridge, United Kingdom and New York, NY, USA.

Bryant, G., Hallett, J., Mason, B., 1960. The epitaxial growth of ice on single-crystalline substrates. J. Phys. Chem. Solids 12, 189–95.

Burton, S. P., Ferrare, R. A., Hostetler, C. A., Hair, J. W., Rogers, R. R., Obland, M. D., Butler, C. F., Cook, A. L., Harper, D. B., Froyd, K. D., 2012. Aerosol classification using airborne high spectral resolution lidar measurements – methodology and examples. Atmos. Meas. Tech. 5 (1), 73–98.

Burton, S. P., Ferrare, R. A., Vaughan, M. A., Omar, A. H., Rogers, R. R., Hostetler, C. A., Hair, J. W., 2013. Aerosol classification from airborne hsrl and comparisons with the calipso vertical feature mask. Atmos. Meas. Tech. 6 (5), 1397–1412.

Cai, Q., Liou, K.-N., 1982. Polarized light scattering by hexagonal ice crystals: theory. Appl. Opt. 21 (19), 3569–3580.

Cai, Y., Montague, D. C., Mooiweer-Bryan, W., Deshler, T., 2008. Performance characteristics of the ultra high sensitivity aerosol spectrometer for particles between 55 and 800nm: Laboratory and field studies. J. Aerosol Sci. 39 (9), 759–769.

Chaikovsky, A., Dubovik, O., Goloub, P., Tanré, D., Pappalardo, G., Wandinger, U., Chaikovskaya, L., Denisov, S., Grudo, Y., Lopatsin, A., et al., 2012. Algorithm and software for the retrieval of vertical aerosol properties using combined lidar/radiometer data: Dissemination in earlinet. In: 26th International Laser Radar Conference, edited by Papayannis, A., Balis, D., and Amiridis, V. pp. 339–402.

Chen, W.-N., Chiang, C.-W., Nee, J.-B., 2002. Lidar ratio and depolarization ratio for cirrus clouds. Appl. Opt. 41 (30), 6470–6476.

Cole, B., Yang, P., Baum, B., Riedi, J., C-Labonnote, L., 2014. Ice particle habit and surface roughness derived from parasol polarization measurements. Atmos. Chem. Phys. 14 (7), 3739–3750.

Cole, B. H., Yang, P., Baum, B. A., Riedi, J., C.-Labonnote, L., Thieuleux, F., Platnick, S., 2013. Comparison of parasol observations with polarized reflectances simulated using different ice habit mixtures. J. Appl. Meteor. Climatol. 52 (1), 186–196.

Connolly, P., Saunders, C., Gallagher, M., Bower, K., Flynn, M., Choularton, T., Whiteway, J., Lawson, R., 2005. Aircraft observations of the influence of electric fields on the aggregation of ice crystals. Q. J. Roy. Meteor. Soc. 131 (608), 1695–1712.

Cotton, R., Osborne, S., Ulanowski, Z., Hirst, E., Kaye, P. H., Greenaway, R., 2010. The ability of the small ice detector (sid-2) to characterize cloud particle and aerosol morphologies obtained during flights of the faam bae-146 research aircraft. J. Armos. Ocean. Tech. 27 (2), 290–303.

Crépel, O., Gayet, J.-F., Fournol, J.-F., Oshchepkov, S., 1997. A new airborne polar nephelometer for the measurement of optical and microphysical cloud properties. part ii: Preliminary tests. In: Annales Geophysicae. Vol. 15. Springer, pp. 460–470.

Dal Maso, M., Hari, P., Kulmala, M., 2009. Spring recovery of photosynthesis and atmospheric particle formation. Boreal Environ. Res. 14 (4), 711–721.

Dal Maso, M., Kulmala, M., Riipinen, I., Wagner, R., Hussein, T., Aalto, P. P., Lehtinen, K. E., 2005. Formation and growth of fresh atmospheric aerosols: eight years of aerosol size distribution data from smear ii, hyytiala, finland. Boreal Environ. Res. 10 (5), 323.

Deierling, W., Petersen, W. A., 2008. Total lightning activity as an indicator of updraft characteristics. J. Geophys. Res. Atmos. 113 (D16).

Del Guasta, M., Niranjan, K., 2001. Observation of low-depolarization contrails at florence (italy) using a 532–1064 nm polarization lidar. Geophys. Res. Lett. 28 (21), 4067–4070.

Del Guasta, M., Vallar, E., Riviere, O., Castagnoli, F., Venturi, V., Morandi, M., 2006. Use of polarimetric lidar for the study of oriented ice plates in clouds. Appl. Opt. 45 (20), 4878–4887.

DeVoe, H., 1964. Optical properties of molecular aggregates. i. classical model of electronic absorption and refraction. J. Chem. Phys. 41 (2), 393–400.

DeVoe, H., 1965. Optical properties of molecular aggregates. ii. classical theory of the refraction, absorption, and optical activity of solutions and crystals. J. Chem. Phys. 43 (9), 3199–3208.

Diedenhoven, B. v., Cairns, B., Fridlind, A., Ackerman, A., Garrett, T. J., 2013. Remote sensing of ice crystal asymmetry parameter using multi-directional polarization measurements–part 2: Application to the research scanning polarimeter. Atmos. Chem. Phys. 13 (6), 3185–3203.

Diedenhoven, B. v., Cairns, B., Geogdzhayev, I., Fridlind, A., Ackerman, A., Yang, P., Baum, B., 2012. Remote sensing of ice crystal asymmetry parameter using multi-directional polarization measurements–part 1: Methodology and evaluation with simulated measurements. Atmos. Meas. Tech. 5 (10), 2361–2374.

Diskin, G. S., Podolske, J. R., Sachse, G. W., Slate, T. A., 2002. Open-path airborne tunable diode laser hygrometer. In: International Symposium on Optical Science and Technology. International Society for Optics and Photonics, pp. 196–204.

Dlugach, J. M., Mishchenko, M. I., 2015. Scattering properties of heterogeneous mineral particles with absorbing inclusions. J. Quant. Spectrosc. Radiat. Transfer 162, 8994.

Dubovik, O., Sinyuk, A., Lapyonok, T., Holben, B. N., Mishchenko, M., Yang, P., Eck, T. F., Volten, H., Munoz, O., Veihelmann, B., et al., 2006. Application of spheroid models to account for aerosol particle nonsphericity in remote sensing of desert dust. J. Geophys. Res. Atmos. 111 (D11).

Duplissy, J., DeCarlo, P. F., Dommen, J., Alfarra, M. R., Metzger, A., Barmpadimos, I., Prevot, A. S., Weingartner, E., Tritscher, T., Gysel, M., et al., 2011. Relating hygroscopicity and composition of organic aerosol particulate matter. Atmos. Chem. Phys. 11 (3), 1155–1165.

Duplissy, J., Enghoff, M. B., Aplin, K. L., Arnold, F., Aufmhoff, H., Avngaard, M., Baltensperger, U., Bondo, T., Bingham, R., Carslaw, K., Curtius, J., David, A., Fastrup, B., Gagné, S., Hahn, F., Harrison, R. G., Kellett, B., Kirkby, J., Kulmala, M., Laakso, L., Laaksonen, A., Lillestol, E., Lockwood, M., Mäkelä,

J., Makhmutov, V., Marsh, N. D., Nieminen, T., Onnela, A., Pedersen, E., Pedersen, J. O. P., Polny, J., Reichl, U., Seinfeld, J. H., Sipilä, M., Stozhkov, Y., Stratmann, F., Svensmark, H., Svensmark, J., Veenhof, R., Verheggen, B., Viisanen, Y., Wagner, P. E., Wehrle, G., Weingartner, E., Wex, H., Wilhelmsson, M., Winkler, P. M., 2010. Results from the cern pilot cloud experiment. Atmos. Chem. Phys. 10 (4), 1635–1647.

Duplissy, J., Merikanto, J., Franchin, A., Tsagkogeorgas, G., Kangasluoma, J., Wimmer, D., Vuollekoski, H., Schobesberger, S., Lehtipalo, K., Flagan, R., et al., 2015. Effect of ions on sulfuric acid-water binary particle formation ii: Experimental data and comparison with qc-normalized classical nucleation theory. J. Geophys. Res. Atmos.

Ebert, M., Inerle-Hof, M., Weinbruch, S., 2002. Environmental scanning electron microscopy as a new technique to determine the hygroscopic behaviour of individual aerosol particles. Atmos. Environ. 36 (39), 5909–5916.

Ehrlich, A., Wendisch, M., Bierwirth, E., Herber, A., Schwarzenböck, A., 2008. Ice crystal shape effects on solar radiative properties of arctic mixed-phase clouds—dependence on microphysical properties. Atmos. Res. 88 (3), 266–276.

Eichler, H., Ehrlich, A., Wendisch, M., Mioche, G., Gayet, J.-F., Wirth, M., Emde, C., Minikin, A., 2009. Influence of ice crystal shape on retrieval of cirrus optical thickness and effective radius: A case study. J. Geophys. Res. Atmos. 114 (D19).

Esselborn, M., Wirth, M., Fix, A., Weinzierl, B., Rasp, K., Tesche, M., Petzold, A., 2009. Spatial distribution and optical properties of saharan dust observed by airborne high spectral resolution lidar during samum 2006. Tellus B 61 (1), 131–143.

Fahey, D., Gao, R., Möhler, O., Saathoff, H., Schiller, C., Ebert, V., Krämer, M., Peter, T., Amarouche, N., Avallone, L., et al., 2014. The aquavit-1 intercomparison of atmospheric water vapor measurement techniques. Atmos. Meas. Tech. 7 (4), 3159–3251.

Febvre, G., Gayet, J.-F., Minikin, A., Schlager, H., Shcherbakov, V., Jourdan, O., Busen, R., Fiebig, M., Kärcher, B., Schumann, U., 2009. On optical and microphysical characteristics of contrails and cirrus. J. Geophys. Res. Atmos. 114 (D2).

Freudenthaler, V., Esselborn, M., Wiegner, M., Heese, B., Tesche, M., Ansmann, A., Müller, D., Althausen, D., Wirth, M., Fix, A., et al., 2009. Depolarization ratio profiling at several wavelengths in pure saharan dust during samum 2006. Tellus B 61 (1), 165–179.

Freudenthaler, V., Homburg, F., Jäger, H., 1995. Contrail observations by ground-based scanning lidar: Cross-sectional growth. Geophys. Res. Lett. 22 (24), 3501–3504.

Frey, W., Borrmann, S., Kunkel, D., Weigel, R., Reus, M. d., Schlager, H., Roiger, A., Voigt, C., Hoor, P., Curtius, J., et al., 2011. In situ measurements of tropical cloud properties in the west african monsoon: upper tropospheric ice clouds, mesoscale convective system outflow, and subvisual cirrus. Atmos. Chem. Phys. 11 (12), 5569–5590.

Fuzzi, S., Andreae, M. O., Huebert, B. J., Kulmala, M., Bond, T. C., Boy, M., Doherty, S. J., Guenther, A., Kanakidou, M., Kawamura, K., Kerminen, V.-M., Lohmann, U., Russell, L. M., Pöschl, U., 2006. Critical assessment of the current state of scientific knowledge, terminology, and research needs concerning the role of organic aerosols in the atmosphere, climate, and global change. Atmos. Chem. Phys. 6 (7), 2017–2038.

Gasteiger, J., Wiegner, M., Groß, S., Freudenthaler, V., Toledano, C., Tesche, M., Kandler, K., 2011. Modelling lidar-relevant optical properties of complex mineral dust aerosols. Tellus B 63 (4), 725–741.

Gayet, J.-F., Asano, S., Yamazaki, A., Uchiyama, A., Sinyuk, A., Jourdan, O., Auriol, F., 2002. Two case studies of winter continental-type water and mixed-phase stratocumuli over the sea 1. microphysical and optical properties. J. Geophys. Res. Atmos. 107 (D21), AAC–11.

Gayet, J.-F., Crépel, O., Fournol, J., Oshchepkov, S., 1997. A new airborne polar nephelometer for the measurements of optical and microphysical cloud properties. part i: Theoretical design. Annales Geophysicae 15 (4), 451–459.

Gayet, J.-F., Mioche, G., Bugliaro, L., Protat, A., Minikin, A., Wirth, M., Dörnbrack, A., Shcherbakov, V., Mayer, B., Garnier, A., et al., 2012a. On the observation of unusual high concentration of small chain-like aggregate ice crystals and large ice water contents near the top of a deep convective cloud during the circle-2 experiment. Atmos. Chem. Phys. 12 (2), 727–744.

Gayet, J.-F., Ovarlez, J., Shcherbakov, V., Ström, J., Schumann, U., Minikin, A., Auriol, F., Petzold, A., Monier, M., 2004. Cirrus cloud microphysical and optical properties at southern and northern midlatitudes during the inca experiment. J. Geophys. Res. Atmos. 109 (D20).

Gayet, J.-F., Shcherbakov, V., Voigt, C., Schumann, U., Schäuble, D., Jessberger, P., Petzold, A., Minikin, A., Schlager, H., Dubovik, O., et al., 2012b. The evolution of microphysical and optical properties of an a380 contrail in the vortex phase. Atmos. Chem. Phys. 12 (14), 6629–6643.

Giangrande, S. E., Collis, S., Straka, J., Protat, A., Williams, C., Krueger, S., 2013. A summary of convective-core vertical velocity properties using arm uhf wind profilers in oklahoma. J. Appl. Meteor. Climatol. 52 (10), 2278–2295.

Groß, S., Esselborn, M., Weinzierl, B., Wirth, M., Fix, A., Petzold, A., 2013. Aerosol classification by airborne high spectral resolution lidar observations. Atmos. Chem. Phys. 13 (5), 2487–2505.

Groß, S., Freudenthaler, V., Schepanski, K., Toledano, C., Schäfler, A., Ansmann, A., Weinzierl, B., 2015. Optical properties of long-range transported saharan dust over barbados as measured by dual-wavelength depolarization raman lidar measurements. Atmos. Chem. Phys. 15 (19), 11067.

Groß, S., Tesche, M., Freudenthaler, V., Toledano, C., Wiegner, M., Ansmann, A., Althausen, D., Seefeldner, M., 2011. Characterization of saharan dust, marine aerosols and mixtures of biomass-burning aerosols and dust by means of multi-wavelength depolarization and raman lidar measurements during samum 2. Tellus B 63 (4), 706–724.

Guenther, A., Hewitt, C. N., Erickson, D., Fall, R., Geron, C., Graedel, T., Harley, P., Klinger, L., Lerdau, M., McKay, W., et al., 1995. A global model of natural volatile organic compound emissions. J. Geophys. Res. Atmos. 100 (D5), 8873–8892.

Hallquist, M., Wenger, J. C., Baltensperger, U., Rudich, Y., Simpson, D., Claeys, M., Dommen, J., Donahue, N. M., George, C., Goldstein, A. H., Hamilton, J. F., Herrmann, H., Hoffmann, T., Iinuma, Y., Jang, M., Jenkin, M. E., Jimenez, J. L., Kiendler-Scharr, A., Maenhaut, W., McFiggans, G., Mentel, T. F., Monod, A., Prévôt, A. S. H., Seinfeld, J. H., Surratt, J. D., Szmigielski, R., Wildt, J., 2009. The formation, properties and impact of secondary organic aerosol: current and emerging issues. Atmos. Chem. Phys. 9 (14), 5155–5236.

Hamed, A., Korhonen, H., Sihto, S.-L., Joutsensaari, J., Järvinen, H., Petäjä, T., Arnold, F., Nieminen, T., Kulmala, M., Smith, J. N., et al., 2011. The role of relative humidity in continental new particle formation. J. Geophys. Res. Atmos. 116 (D3).

Haner, D. A., McGuckin, B. T., Bruegge, C. J., 1999. Polarization characteristics of spectralon illuminated by coherent light. Appl. Opt. 38 (30), 6350–6356.

Haralick, R. M., Shanmugam, K., Dinstein, I. H., 1973. Textural features for image classification. IEEE Trans. Syst., Man, Cybern., Syst. (6), 610–621.

Hayman, M., Spuler, S., Morley, B., Joseph, V., 2012. Polarization lidar operation for measuring backscatter phase matrices of oriented scatterers. Opt. Express 20 (28), 29553–29567.

Heintzenberg, J., 2009. The samum-1 experiment over southern morocco: Overview and introduction. Tellus B 61 (1), 2–11.

Heymsfield, A. J., Aulenbach, S. M., 1998. Shapes, sizes and light scattering properties of ice crystals in cirrus and a persistent contrail during success. Geophys. Res. Lett. 25 (9), 1331–1334.

Heymsfield, A. J., Miloshevich, L. M., Schmitt, C., Bansemer, A., Twohy, C., Poellot, M. R., Fridlind, A., Gerber, H., 2005. Homogeneous ice nucleation in subtropical and tropical convection and its influence on cirrus anvil microphysics. J. Atmos. Sci. 62 (1), 41–64.

Heymsfield, A. J., Platt, C., 1984. A parameterization of the particle size spectrum of ice clouds in terms of the ambient temperature and the ice water content. J. Atmos. Sci. 41 (5), 846–855.

Heymsfield, A. J., Sabin, R. M., 1989. Cirrus crystal nucleation by homogeneous freezing of solution droplets. J. Atmos. Sci. 46 (14), 2252–2264.

Heymsfield, A. J., Schmitt, C., Bansemer, A., Twohy, C. H., 2010. Improved representation of ice particle masses based on observations in natural clouds. J. Atmos. Sci. 67 (10), 3303–3318.

Hinds, W. C., 2012. Aerosol technology: properties, behavior, and measurement of airborne particles. John Wiley & Sons.

Hiranuma, N., Paukert, M., Steinke, I., Zhang, K., Kulkarni, G., Hoose, C., Schnaiter, M., Saathoff, H., Möhler, O., 2014. A comprehensive parameterization of heterogeneous ice nucleation of dust surrogate: laboratory study with hematite particles and its application to atmospheric models. Atmos. Chem. Phys. 14 (23), 13145–13158.

Hirsikko, A., O'Connor, E., Komppula, M., Korhonen, K., Pfüller, A., Giannakaki, E., Wood, C., Bauer-Pfundstein, M., Poikonen, A., Karppinen, T., et al., 2014. Observing wind, aerosol particles, cloud and precipitation: Finland's new ground-based remote-sensing network. Atmos. Meas. Tech. 7 (5), 1351–1375.

Hovenier, J., Guirado, D., 2014. Zero slopes of the scattering function and scattering matrix for strict forward and backward scattering by mirror symmetric collections of randomly oriented particles. J. Quant. Spectrosc. Radiat. Transfer 133, 596–602.

Hovenier, J., Volten, H., Munoz, O., Van der Zande, W., Waters, L., 2003. Laboratory studies of scattering matrices for randomly oriented particles: potentials, problems, and perspectives. J. Quant. Spectrosc. Radiat. Transfer 79, 741–755.

Hoyle, C., Boy, M., Donahue, N., Fry, J., Glasius, M., Guenther, A., Hallar, A., Huff Hartz, K., Petters, M., Petäjä, T., et al., 2011. A review of the anthropogenic influence on biogenic secondary organic aerosol. Atmos. Chem. Phys. 11 (1), 321–343.

Hoyle, C. R., Fuchs, C., Järvinen, E., Saathoff, H., Dias, A., El Haddad, I., Gysel, M., Coburn, S. C., Tröstl, J., Bernhammer, A.-K., Bianchi, F., Breitenlechner, M., Corbin, J. C., Craven, J., Donahue, N. M., Duplissy, J., Ehrhart, S., Frege, C., Gordon, H., Höppel, N., Heinritzi, M., Kristensen, T. B., Molteni, U., Nichman, L., Pinterich, T., Prévôt, A. S. H., Simon, M., Slowik, J. G., Steiner, G., Tomé, A., Vogel, A. L., Volkamer, R., Wagner, A. C., Wagner, R., Wexler, A. S., Williamson, C., Winkler, P. M., Yan, C., Amorim, A., Dommen, J., Curtius, J., Gallagher, M. W., Flagan, R. C., Hansel, A., Kirkby, J., Kulmala, M., Möhler, O., Stratmann, F., Worsnop, D. R., Baltensperger, U., 2016. Aqueous phase oxidation of sulphur dioxide by ozone in cloud droplets. Atmos. Chem. Phys. 16 (3), 1693–1712.

Huang, R.-J., Zhang, Y., Bozzetti, C., Ho, K.-F., Cao, J.-J., Han, Y., Daellenbach, K. R., Slowik, J. G., Platt, S. M., Canonaco, F., et al., 2014. High secondary aerosol contribution to particulate pollution during haze events in china. Nature 514 (7521), 218–222.

Ignatius, K., Kristensen, T., Järvinen, E., Nichman, L., Fuchs, C., Gordon, H., Herenz, P., Hoyle, C., Duplissy, J., Garimella, S., et al., 2015. Heterogeneous ice nucleation of viscous secondary organic aerosol produced from ozonolysis of α-pinene. Atmos. Chem. Phys. Discuss. 15 (24), 35719–35752.

Illingworth, A., Barker, H., Beljaars, A., Ceccaldi, M., Chepfer, H., Cole, J., Delanoë, J., Domenech, C., Donovan, D., Fukuda, S., et al., 2014. The earthcare satellite: The next step forward in global measurements of clouds, aerosols, precipitation and radiation. Bull. Amer. Meteor. Soc.

Ishimoto, H., Zaizen, Y., Uchiyama, A., Masuda, K., Mano, Y., 2010. Shape modeling of mineral dust particles for light-scattering calculations using the spatial poisson–voronoi tessellation. J. Quant. Spectrosc. Radiat. Transfer 111 (16), 2434–2443.

Järvinen, E., Ignatius, K., Nichman, L., Kristensen, T. B., Fuchs, C., Hoyle, C. R., Höppel, N., Corbin, J. C., Craven, J., Duplissy, J., Ehrhart, S., El Haddad, I., Frege, C., Gordon, H., Jokinen, T., Kallinger, P., Kirkby, J., Kiselev, A., Naumann, K.-H., Petäjä, T., Pinterich, T., Prevot, A. S. H., Saathoff, H., Schiebel, T., Sengupta, K., Simon, M., Slowik, J. G., Tröstl, J., Virtanen, A., Vochezer, P., Vogt, S., Wagner, A. C., Wagner, R., Williamson, C., Winkler, P. M., Yan, C., Baltensperger, U., Donahue, N. M., Flagan, R. C., Gallagher, M., Hansel, A., Kulmala, M., Stratmann, F., Worsnop, D. R., Möhler, O., Leisner, T., Schnaiter, M., 2016b. Observation of viscosity transition in -pinene secondary organic aerosol. Atmos. Chem. Phys. 16 (7), 4423–4438.

Järvinen, E., Kemppinen, O., Nousiainen, T., Kociok, T., Möhler, O., Leisner, T., Schnaiter, M., 2016a. Laboratory investigations of mineral dust near-backscattering depolarization ratios. J. Quant. Spectrosc. Radiat. Transfer.

Järvinen, E., Schnaiter, M., Mioche, G., Jourdan, O., Shcherbakov, V. N., Costa, A., Afchine, A., Krämer, M., Heidelberg, F., Jurkat, T., Voigt, C., Schlager, H., Nichman, L., Gallagher, M., Hirst, E., Schmitt, C. G., Bansemer, A., Heymsfield, A. J., Lawson, P., Tricoli, U., Pfeilsticker, K., Vochezer, P., Möhler, O.,

Leisner, T., 2016c. Near-spherical ice in convective clouds. J. Atmos. Sci. accepted.

Järvinen, E., Vochezer, P., Möhler, O., Schnaiter, M., 2014. Laboratory study of microphysical and scattering properties of corona-producing cirrus clouds. Appl. Opt. 53 (31), 7566–7575.

Jensen, E. J., Toon, O. B., Selkirk, H. B., Spinhirne, J. D., Schoeberl, M. R., 1996. On the formation and persistence of subvisible cirrus clouds near the tropical tropopause. J. Geophys. Res. Atmos. 101 (D16), 21361–21375.

Jimenez, J. L., Canagaratna, M. R., Donahue, N. M., Prevot, A. S. H., Zhang, Q., Kroll, J. H., DeCarlo, P. F., Allan, J. D., Coe, H., Ng, N. L., Aiken, A. C., Docherty, K. S., Ulbrich, I. M., Grieshop, A. P., Robinson, A. L., Duplissy, J., Smith, J. D., Wilson, K. R., Lanz, V. A., Hueglin, C., Sun, Y. L., Tian, J., Laaksonen, A., Raatikainen, T., Rautiainen, J., Vaattovaara, P., Ehn, M., Kulmala, M., Tomlinson, J. M., Collins, D. R., Cubison, M. J., E., Dunlea, J., Huffman, J. A., Onasch, T. B., Alfarra, M. R., Williams, P. I., Bower, K., Kondo, Y., Schneider, J., Drewnick, F., Borrmann, S., Weimer, S., Demerjian, K., Salcedo, D., Cottrell, L., Griffin, R., Takami, A., Miyoshi, T., Hatakeyama, S., Shimono, A., Sun, J. Y., Zhang, Y. M., Dzepina, K., Kimmel, J. R., Sueper, D., Jayne, J. T., Herndon, S. C., Trimborn, A. M., Williams, L. R., Wood, E. C., Middlebrook, A. M., Kolb, C. E., Baltensperger, U., Worsnop, D. R., 2009. Evolution of organic aerosols in the atmosphere. Science 326 (5959), 1525–1529.

Johnson, A., Lasher-Trapp, S., Bansemer, A., Ulanowski, Z., Heymsfield, A. J., 2014. Difficulties in early ice detection with the small ice detector-2 hiaper (sid-2h) in maritime cumuli. J. Armos. Ocean. Tech. 31 (6), 1263–1275.

Jourdan, O., Mioche, G., Garrett, T. J., Schwarzenböck, A., Vidot, J., Xie, Y., Shcherbakov, V., Yang, P., Gayet, J.-F., 2010. Coupling of the microphysical and optical properties of an arctic nimbostratus cloud during the astar 2004 experiment: Implications for light-scattering modeling. J. Geophys. Res. Atmos. 115 (D23).

Jourdan, O., Oshchepkov, S., Shcherbakov, V., Gayet, J.-F., Isaka, H., 2003. Assessment of cloud optical parameters in the solar region: Retrievals from airborne measurements of scattering phase functions. J. Geophys. Res. Atmos. 108 (D18).

Kahnert, M., Nousiainen, T., Veihelmann, B., 2005. Spherical and spheroidal model particles as an error source in aerosol climate forcing and radiance computations: A case study for feldspar aerosols. J. Geophys. Res. Atmos. 110 (D18).

Kaller, W., 2004. A new polar nephelometer for measurement of atmospheric aerosols. J. Quant. Spectrosc. Radiat. Transfer 87 (2), 107–117.

Kärcher, B., Yu, F., 2009. Role of aircraft soot emissions in contrail formation. Geophys. Res. Lett. 36 (1).

Kaye, P. H., Hirst, E., Greenaway, R. S., Ulanowski, Z., Hesse, E., DeMott, P. J., Saunders, C., Connolly, P., 2008. Classifying atmospheric ice crystals by spatial light scattering. Opt. Lett. 33 (13), 1545–1547.

Kemppinen, O., Nousiainen, T., Jeong, G., 2015a. Effects of dust particle internal structure on light scattering. Atmos. Chem. Phys. Discuss. 15 (14), 20349–20394.

Kemppinen, O., Nousiainen, T., Lindqvist, H., 2015b. The impact of surface roughness on scattering by realistically shaped wavelength-scale dust particles. J. Quant. Spectrosc. Radiat. Transfer 150, 55–67.

Kemppinen, O., Nousiainen, T., Merikallio, S., Räisänen, P., 2015c. Retrieving microphysical properties of dust-like particles using ellipsoids: the case of refractive index. Atmos. Chem. Phys. 15, 11117–11132.

Kidd, C., Perraud, V., Wingen, L. M., Finlayson-Pitts, B. J., 2014. Integrating phase and composition of secondary organic aerosol from the ozonolysis of α-pinene. Proc. Nat. Acad. Sci. USA 111 (21), 7552–7557.

Kim, H., Barkey, B., Paulson, S. E., 2010. Real refractive indices of α-and β-pinene and toluene secondary organic aerosols generated from ozonolysis and photo-oxidation. J. Geophys. Res. Atmos. 115 (D24).

Kirkby, J., Curtius, J., Almeida, J., Dunne, E., Duplissy, J., Ehrhart, S., Franchin, A., Gagné, S., Ickes, L., Kürten, A., et al., 2011. Role of sulphuric acid, ammonia and galactic cosmic rays in atmospheric aerosol nucleation. Nature 476 (7361), 429–433.

Koop, T., Bookhold, J., Shiraiwa, M., Pöschl, U., 2011. Glass transition and phase state of organic compounds: dependency on molecular properties and implications for secondary organic aerosols in the atmosphere. Phys. Chem. Chem. Phys. 13 (43), 19238–19255.

Korolev, A., Isaac, G., 2003. Roundness and aspect ratio of particles in ice clouds. J. Atmos. Sci. 60 (15), 1795–1808.

Korolev, A., Isaac, G., Hallett, J., 1999. Ice particle habits in arctic clouds. Geophys. Res. Lett. 26 (9), 1299–1302.

Krämer, M., Schiller, C., Afchine, A., Bauer, R., Gensch, I., Mangold, A., Schlicht, S., Spelten, N., Sitnikov, N., Borrmann, S., et al., 2009. Ice supersaturations and cirrus cloud crystal numbers. Atmos. Chem. Phys. 9 (11), 3505–3522.

Kulmala, M., Kontkanen, J., Junninen, H., Lehtipalo, K., Manninen, H. E., Nieminen, T., Petäjä, T., Sipilä, M., Schobesberger, S., Rantala, P., et al., 2013. Direct observations of atmospheric aerosol nucleation. Science 339 (6122), 943–946.

Kulmala, M., Riipinen, I., Sipilä, M., Manninen, H. E., Petäjä, T., Junninen, H., Dal Maso, M., Mordas, G., Mirme, A., Vana, M., et al., 2007. Toward direct measurement of atmospheric nucleation. Science 318 (5847), 89–92.

Kupc, A., Amorim, A., Curtius, J., Danielczok, A., Duplissy, J., Ehrhart, S., Walther, H., Ickes, L., Kirkby, J., Kürten, A., et al., 2011. A fibre-optic uv system for h 2 so 4 production in aerosol chambers causing minimal thermal effects. J. Aerosol Sci. 42 (8), 532–543.

Laakso, L., Hussein, T., Aarnio, P., Komppula, M., Hiltunen, V., Viisanen, Y., Kulmala, M., 2003. Diurnal and annual characteristics of particle mass and number concentrations in urban, rural and arctic environments in finland. Atmos. Environ. 37 (19), 2629–2641.

Laaksonen, A., Kulmala, M., O'Dowd, C. D., Joutsensaari, J., Vaattovaara, P., Mikkonen, S., Lehtinen, K. E. J., Sogacheva, L., Dal Maso, M., Aalto, P., Petäjä, T., Sogachev, A., Yoon, Y. J., Lihavainen, H., Nilsson, D., Facchini, M. C., Cavalli, F., Fuzzi, S., Hoffmann, T., Arnold, F., Hanke, M., Sellegri, K., Umann, B., Junkermann, W., Coe, H., Allan, J. D., Alfarra, M. R., Worsnop, D. R., Riekkola, M. L., Hyötyläinen, T., Viisanen, Y., 2008. The role of voc oxidation products in continental new particle formation. Atmos. Chem. Phys. 8 (10), 2657–2665.

Ladino, L. A., Zhou, S., Yakobi-Hancock, J. D., Aljawhary, D., Abbatt, J. P. D., 2014. Factors controlling the ice nucleating abilities of α-pinene soa particles. J. Geophys. Res. Atmos. 119, 9041–9051.

Lambe, A., Onasch, T., Massoli, P., Croasdale, D., Wright, J., Ahern, A., Williams, L., Worsnop, D., Brune, W., Davidovits, P., 2011. Laboratory studies of the chemical composition and cloud condensation nuclei (ccn) activity of secondary organic aerosol (soa) and oxidized primary organic aerosol (opoa). Atmos. Chem. Phys. 11 (17), 8913–8928.

Lampert, A., Ehrlich, A., Dörnbrack, A., Jourdan, O., Gayet, J.-F., Mioche, G., Shcherbakov, V., Ritter, C., Wendisch, M., 2009. Microphysical and radiative characterization of a subvisible midlevel arctic ice cloud by airborne observations–a case study. Atmos. Chem. Phys. 9 (8), 2647–2661.

Lawson, R. P., Baker, B., Pilson, B., Mo, Q., 2006. In situ observations of the microphysical properties of wave, cirrus, and anvil clouds. part ii: Cirrus clouds. J. Atmos. Sci. 63 (12), 3186–3203.

Lawson, R. P., Baker, B. A., Pilson, B. L., 2003. In-situ measurements of microphysical properties of mid-latitude and anvil cirrus. In: Proceedings, 30th International Symposium on Remote Sensing of Environment, Honolulu, Hawaii, November. pp. 707–710.

Lawson, R. P., Jensen, E., Mitchell, D. L., Baker, B., Mo, Q., Pilson, B., 2010. Microphysical and radiative properties of tropical clouds investigated in tc4 and namma. J. Geophys. Res. Atmos. 115 (D10).

Li, X., Taflove, A., Backman, V., 2005. Quantitative analysis of depolarization of backscattered light by stochastically inhomogeneous dielectric particles. Opt. Lett. 30 (8), 902–904.

Lienhard, D., Huisman, A., Krieger, U., Rudich, Y., Marcolli, C., Luo, B., Bones, D., Reid, J., Lambe, A., Canagaratna, M., et al., 2015. Viscous organic aerosol particles in the upper troposphere: diffusivity-controlled water uptake and ice nucleation? Atmos. Chem. Phys. 15 (23), 13599–13613.

Lindqvist, H., Jokinen, O., Kandler, K., Nousiainen, T., 2011. Modelling mineral dust using stereophotogrammetry. AAPP— Physical, Mathematical, and Natural Sciences 89 (S1).

Lindqvist, H., Jokinen, O., Kandler, K., Scheuvens, D., Nousiainen, T., 2014. Single scattering by realistic, inhomogeneous mineral dust particles with stereogrammetric shapes. Atmos. Chem. Phys. 14 (1), 143–157.

Linke, C., Möhler, O., Veres, A., Mohacsi, A., Bozóki, Z., Szabó, G., Schnaiter, M., 2006. Optical properties and mineralogical composition of different saharan mineral dust samples: a laboratory study. Atmos. Chem. Phys. 6 (2), 2897–2922.

Liu, C., Panetta, R. L., Yang, P., Macke, A., Baran, A. J., 2013. Modeling the scattering properties of mineral aerosols using concave fractal polyhedra. Appl. Opt. 52 (4), 640–652.

Liu, C., Yang, P., Minnis, P., Loeb, N., Kato, S., Heymsfield, A., Schmitt, C., 2014. A two-habit model for the microphysical and optical properties of ice clouds. Atmos. Chem. Phys. 14 (24), 13719–13737.

Liu, L., Mishchenko, M. I., Hovenier, J. W., Volten, H., Muñoz, O., 2003. Scattering matrix of quartz aerosols: comparison and synthesis of laboratory and lorenz–mie results. J. Quant. Spectrosc. Radiat. Transfer 79, 911–920.

Liu, Z., Omar, A., Vaughan, M., Hair, J., Kittaka, C., Hu, Y., Powell, K., Trepte, C., Winker, D., Hostetler, C., 2008. CALIPSO lidar observations of the optical properties of saharan dust: A case study of long range transport. J. Geophys. Res. Atmos. 113 (D7).

Lu, R.-S., Tian, G.-Y., Gledhill, D., Ward, S., 2006. Grinding surface roughness measurement based on the co-occurrence matrix of speckle pattern texture. Appl. Opt. 45 (35), 8839–8847.

Luebke, A., Afchine, A., Costa, A., Meyer, J., Rolf, C., Spelten, N., Avallone, L., Baumgardner, D., Krämer, M., 2015. The origin of midlatitude ice clouds and the resulting influence on their microphysical properties. Atmos. Chem. Phys. Discuss. 15 (23), 34243–34281.

Macke, A., 1993. Scattering of light by polyhedral ice crystals. Appl. Opt. 32 (15), 2780–2788.

Magee, N., Miller, A., Amaral, M., Cumiskey, A., 2014. Mesoscopic surface roughness of ice crystals pervasive across a wide range of ice crystal conditions. Atmos. Chem. Phys. 14 (22), 12357–12371.

Mamouri, R., Ansmann, A., 2014. Fine and coarse dust separation with polarization lidar. Atmos. Meas. Tech. 7 (11), 3717–3735.

Marshall, T. C., Rust, W. D., Stolzenburg, M., 1995. Electrical structure and updraft speeds in thunderstorms over the southern great plains. J. Geophys. Res. Atmos. 100 (D1), 1001–1015.

Martins, E., Noel, V., Chepfer, H., 2011. Properties of cirrus and subvisible cirrus from nighttime cloud-aerosol lidar with orthogonal polarization (caliop), related to atmospheric dynamics and water vapor. J. Geophys. Res. Atmos. 116 (D2).

Massoli, P., Lambe, A., Ahern, A., Williams, L., Ehn, M., Mikkilä, J., Canagaratna, M., Brune, W., Onasch, T., Jayne, J., et al., 2010. Relationship between aerosol oxidation level and hygroscopic properties of laboratory generated secondary organic aerosol (soa) particles. Geophys. Res. Lett. 37 (24).

Mauno, P., McFarquhar, G. M., Räisänen, P., Kahnert, M., Timlin, M. S., Nousiainen, T., 2011. The influence of observed cirrus microphysical properties on shortwave radiation: A case study over oklahoma. J. Geophys. Res. Atmos. 116 (D22).

May, P. T., Mather, J. H., Vaughan, G., Jakob, C., McFarquhar, G. M., Bower, K. N., Mace, G. G., 2008. The tropical warm pool international cloud experiment. Bull. Amer. Meteor. Soc. 89 (5), 629.

McFarquhar, G. M., Heymsfield, A. J., 1996. Microphysical characteristics of three anvils sampled during the central equatorial pacific experiment. J. Atmos. Sci. 53 (17), 2401–2423.

McFarquhar, G. M., Um, J., Jackson, R., 2013. Small cloud particle shapes in mixed-phase clouds. Journal of Applied Meteorology and Climatology 52 (5), 1277–1293.

Merikallio, S., Lindqvist, H., Nousiainen, T., Kahnert, M., 2011. Modelling light scattering by mineral dust using spheroids: assessment of applicability. Atmos. Chem. Phys. 11 (11), 5347–5363.

Meyer, J., 2012. Ice crystal measurements with the new particle spectrometer nixe-caps. Ph.D. thesis.

Mie, G., 1908. Beiträge zur optik trüber medien, speziell kolloidaler metallösungen. Annalen der physik 330 (3), 377–445.

Miffre, A., Mehri, T., Francis, M., Rairoux, P., 2016. Uv–vis depolarization from arizona test dust particles at exact backscattering angle. J. Quant. Spectrosc. Radiat. Transfer 169, 79–90.

Mikhailov, E., Vlasenko, S., Martin, S., Koop, T., Pöschl, U., 2009. Amorphous and crystalline aerosol particles interacting with water vapor: conceptual framework and experimental evidence for restructuring, phase transitions and kinetic limitations. Atmos. Chem. Phys. 9 (24), 9491–9522.

Minutolo, P., Gambi, G., D'ALESSIO, A., D'ANNA, A., 1994. Optical and spectroscopic characterization of rich premixed flames across the soot formation threshold. Combust. Sci. Technol. 101 (1-6), 311–325.

Mishchenko, M., Hovenier, J., 1995b. Depolarization of light backscattered by randomly oriented nonspherical particles. Opt. Lett. 20 (12), 1356–1358.

Mishchenko, M., Lacis, A., Carlson, B., Travis, L., 1995a. Nonsphericity of dust-like tropospheric aerosols: Implications for aerosol remote sensing and climate modeling. Geophys. Res. Lett 22 (9), 1077–1080.

Mishchenko, M., Travis, L., Mackowski, D., 1996. T-matrix computations of light scattering by nonspherical particles: a review. J. Quant. Spectrosc. Radiat. Transfer 55 (5), 535–575.

Mishchenko, M. I., 2009. Electromagnetic scattering by nonspherical particles: A tutorial review. J. Quant. Spectrosc. Radiat. Transfer 110 (11), 808–832.

Mishchenko, M. I., Sassen, K., 1998. Depolarization of lidar returns by small ice crystals: An application to contrails. Geophys. Res. Lett. 25 (3), 309–312.

Mishchenko, M. I., Travis, L. D., 1997. Satellite retrieval of aerosol properties over the ocean using polarization as well as intensity of reflected sunlight. J. Geophys. Res. Atmos. 102 (D14), 16989–17013.

Mitchell, D. L., Rasch, P., Ivanova, D., McFarquhar, G., Nousiainen, T., 2008. Impact of small ice crystal assumptions on ice sedimentation rates in cirrus clouds and gcm simulations. Geophys. Res. Lett. 35 (9).

Möhler, O., Büttner, S., Linke, C., Schnaiter, M., Saathoff, H., Stetzer, O., Wagner, R., Krämer, M., Mangold, A., Ebert, V., et al., 2005. Effect of sulfuric acid coating on heterogeneous ice nucleation by soot aerosol particles. J. Geophys. Res. Atmos. 110 (D11).

Möhler, O., Georgakopoulos, D., Morris, C., Benz, S., Ebert, V., Hunsmann, S., Saathoff, H., Schnaiter, M., Wagner, R., 2008. Heterogeneous ice nucleation activity of bacteria: new laboratory experiments at simulated cloud conditions. Biogeosciences 5 (5), 1425–1435.

Möhler, O., Stetzer, O., Schaefers, S., Linke, C., Schnaiter, M., Tiede, R., Saathoff, H., Krämer, M., Mangold, A., Budz, P., et al., 2003. Experimental investigation of homogeneous freezing of sulphuric acid particles in the aerosol chamber aida. Atmos. Chem. Phys. 3 (1), 211–223.

Morgan, M. A., Mei, K. K., 1979. Finite-element computation of scattering by inhomogeneous penetrable bodies of revolution. IEEE Trans. Antennas Propag. 27 (2), 202–214.

Müller, D., Veselovskii, I., Kolgotin, A., Tesche, M., Ansmann, A., Dubovik, O., 2013. Vertical profiles of pure dust and mixed smoke–dust plumes inferred from inversion of multiwavelength raman/polarization lidar data and comparison to aeronet retrievals and in situ observations. Appl. Opt. 52 (14), 3178–3202.

Muñoz, O., Moreno, F., Guirado, D., Ramos, J., López, A., Girela, F., Jerónimo, J., Costillo, L., Bustamante, I., 2010. Experimental determination of scattering matrices of dust particles at visible wavelengths: The iaa light scattering apparatus. J. Quant. Spectrosc. Radiat. Transfer 111 (1), 187–196.

Muñoz, O., Volten, H., Hovenier, J., Nousiainen, T., Muinonen, K., Guirado, D., Moreno, F., Waters, L., 2007. Scattering matrix of large saharan dust particles: experiments and computations. J. Geophys. Res. Atmos. 112 (D13).

Murphy, D., Koop, T., 2005. Review of the vapour pressures of ice and supercooled water for atmospheric applications. Q. J. Roy. Meteor. Soc. 131 (608), 1539–1565.

Murray, B. J., Wilson, T. W., Dobbie, S., Cui, Z., Al-Jumur, S. M., Möhler, O., Schnaiter, M., Wagner, R., Benz, S., Niemand, M., et al., 2010. Heterogeneous nucleation of ice particles on glassy aerosols under cirrus conditions. Nat. Geosci. 3 (4), 233–237.

Naumann, K.-H., 2003. Cosima—a computer program simulating the dynamics of fractal aerosols. J. Aerosol Sci. 34 (10), 1371 - 1397.

Nazaryan, H., McCormick, M. P., Menzel, W. P., 2008. Global characterization of cirrus clouds using calipso data. J. Geophys. Res. Atmos. 113 (D16).

Nel, A., 2005. Air pollution-related illness: Effects of particles. Science 308 (5723), 804–806.

Neshyba, S., Lowen, B., Benning, M., Lawson, A., Rowe, P., 2013. Roughness metrics of prismatic facets of ice. J. Geophys. Res. Atmos. 118 (8), 3309–3318.

Nichman, L., Fuchs, C., Järvinen, E., Ignatius, K., Höppel, N. F., Dias, A., Heinritzi, M., Simon, M., Tröstl, J., Wagner, A. C., Wagner, R., Williamson, C., Yan, C., Connolly, P. J., Dorsey, J. R., Duplissy, J., Ehrhart, S., Frege, C., Gordon, H., Hoyle, C. R., Kristensen, T. B., Steiner, G., McPherson Donahue, N., Flagan, R., Gallagher, M. W., Kirkby, J., Möhler, O., Saathoff, H., Schnaiter, M., Stratmann, F., Tomé, A., 2016. Phase transition observations and discrimination of small cloud particles by light polarization in expansion chamber experiments. Atmos. Chem. Phys 16 (5), 3651–3664.

Noel, V., Winker, D. M., McGill, M., Lawson, P., 2004. Classification of particle shapes from lidar depolarization ratio in convective ice clouds compared to in situ observations during crystal-face. J. Geophys. Res. Atmos. 109 (D24).

Nousiainen, T., 2009. Optical modeling of mineral dust particles: A review. J. Quant. Spectrosc. Radiat. Transfer 110 (14-16), 1261–1279.

Nousiainen, T., Kahnert, M., Veihelmann, B., 2006. Light scattering modeling of small feldspar aerosol particles using polyhedral prisms and spheroids. J. Quant. Spectrosc. Radiat. Transfer 101 (3), 471–487.

Nousiainen, T., McFarquhar, G. M., 2004. Light scattering by quasi-spherical ice crystals. J. Atmos. Sci. 61 (18), 2229–2248.

Nousiainen, T., Zubko, E., Lindqvist, H., Kahnert, M., Tyynelä, J., 2012. Comparison of scattering by different nonspherical, wavelength-scale particles. J. Quant. Spectrosc. Radiat. Transfer 113 (18), 2391–2405.

Oguchi, T., Hosoya, Y., 1974. Scattering properties of oblate raindrops and cross polarization of radio waves due to rain. ii-calculations at microwave and millimeter wave regions. J. Radio Res. Lab. 21, 191–259.

Olsen, A. P., Flagan, R. C., Kornfield, J. A., 2006. Manipulation of athermal nuclei in aqueous poly(ethylene oxide) by scanning activity gravimetric analysis. Macromolecules 39 (24), 84198427.

Omar, A. H., Winker, D. M., Vaughan, M. A., Hu, Y., Trepte, C. R., Ferrare, R. A., Lee, K.-P., Hostetler, C. A., Kittaka, C., Rogers, R. R., et al., 2009. The calipso automated aerosol classification and lidar ratio selection algorithm. J. Armos. Ocean. Tech. 26 (10), 1994–2014.

Ono, A., 1969. The shape and riming properties of ice crystals in natural clouds. J. Atmos. Sci. 26 (1), 138–147.

Pajunoja, A., Lambe, A. T., Hakala, J., Rastak, N., Cummings, M. J., Brogan, J. F., Hao, L., Paramonov, M., Hong, J., Prisle, N. L., et al., 2015. Adsorptive uptake of water by semisolid secondary organic aerosols. Geophys. Res. Lett. 42 (8), 3063–3068.

Pajunoja, A., Malila, J., Hao, L., Joutsensaari, J., Lehtinen, K. E., Virtanen, A., 2014. Estimating the viscosity range of soa particles based on their coalescence time. Aerosol Sci. Tech. 48 (2), i–iv.

Philips, V. T., Donner, L. J., Garner, S. T., 2007. Nucleation processes in deep convection simulated by a cloud-system-resolving model with double-moment bulk microphysics. J. Atmos. Sci. 64 (3), 738–761.

Plass, G., Kattawar, G., 1971. Radiative transfer in water and ice clouds in the visible and infrared region. Appl. Opt. 10 (4), 738–748.

Platnick, S., King, M. D., Ackerman, S. A., Menzel, W. P., Baum, B. A., Riédi, J. C., Frey, R. A., 2003. The modis cloud products: Algorithms and examples from terra. IEEE Trans. Geosci. Remote Sens. 41 (2), 459–473.

Pregla, R., 1989. The method of lines. Analysis of Electromagnetic Fields and Waves: The Method of Lines, 1–13.

Price, H. C., Mattsson, J., Zhang, Y., Bertram, A. K., Davies, J. F., Grayson, J. W., Martin, S. T., O'Sullivan, D., Reid, J. P., Rickards, A. M., et al., 2015. Water diffusion in atmospherically relevant α-pinene secondary organic material. Chem. Sci. 6 (8), 4876–4883.

Prospero, J. M., Ginoux, P., Torres, O., Nicholson, S. E., Gill, T. E., 2002. Environmental characterization of global sources of atmospheric soil dust identified with the nimbus 7 total ozone mapping spectrometer (toms) absorbing aerosol product. Rev. Geophys. 40 (1), 2–1.

Purcell, E. M., Pennypacker, C. R., 1973. Scattering and absorption of light by nonspherical dielectric grains. Astrophys. J. 186, 705–714.

Ramaswamy, V., Boucher, O., Haigh, J., Hauglustine, D., Haywood, J., Myhre, G., Nakajima, T., Shi, G., Solomon, S., 2001. Radiative forcing of climate. Clim. Change, 349–416.

Renbaum-Wolff, L., Grayson, J. W., Bateman, A. P., Kuwata, M., Sellier, M., Murray, B. J., Shilling, J. E., Martin, S. T., Bertram, A. K., 2013. Viscosity of α-pinene secondary organic material and implications for particle growth and reactivity. Proc. Nat. Acad. Sci. USA 110 (20), 8014–8019.

Reus, M. d., Borrmann, S., Bansemer, A., Heymsfield, A., Weigel, R., Schiller, C., Mitev, V., Frey, W., Kunkel, D., Kürten, A., et al., 2009. Evidence for ice particles in the tropical stratosphere from in-situ measurements. Atmos. Chem. Phys. 9 (18), 6775–6792.

Riccobono, F., Schobesberger, S., Scott, C. E., Dommen, J., Ortega, I. K., Rondo, L., Almeida, J., Amorim, A., Bianchi, F., Breitenlechner, M., et al., 2014. Oxidation products of biogenic emissions contribute to nucleation of atmospheric particles. Science 344 (6185), 717–721.

Riipinen, I., Yli-Juuti, T., Pierce, J. R., Petäjä, T., Worsnop, D. R., Kulmala, M., Donahue, N. M., 2012. The contribution of organics to atmospheric nanoparticle growth. Nat. Geosci. 5 (7), 453–458.

Rosenfeld, D., Woodley, W. L., 2000. Deep convective clouds with sustained supercooled liquid water down to-37.5 c. Nature 405 (6785), 440–442.

Sakai, T., Nagai, T., Zaizen, Y., Mano, Y., 2010. Backscattering linear depolarization ratio measurements of mineral, sea-salt, and ammonium sulfate particles simulated in a laboratory chamber. Appl. Opt. 49 (23), 4441–4449.

Sakai, T., Shibata, T., Kwon, S.-A., Kim, Y.-S., Tamura, K., Iwasaka, Y., 2000. Free tropospheric aerosol backscatter, depolarization ratio, and relative humidity measured with the raman lidar at nagoya in 1994–1997: contributions of aerosols from the asian continent and the pacific ocean. Atmospheric Environment 34 (3), 431–442.

Sassen, K., 1991. The polarization lidar technique for cloud research: A review and current assessment. Bull. Amer. Meteor. Soc. 72 (12), 1848–1866.

Saukko, E., Lambe, A. T., Massoli, P., Koop, T., Wright, J. P., Croasdale, D. R., Pedernera, D. A., Onasch, T. B., Laaksonen, A., Davidovits, P., Worsnop, D. R., Virtanen, A., 2012. Humidity-dependent phase state of soa particles from biogenic and anthropogenic precursors. Atmos. Chem. Phys. 12 (16), 7517–7529.

Sazaki, G., Zepeda, S., Nakatsubo, S., 2010. Elementary steps at the surface of ice crystals visualized by advanced optical microscopy. Proc. Nat. Acad. Sci. USA 107 (19702–19707).

Schill, G. P., De Haan, D. O., Tolbert, M. A., 2014. Heterogeneous ice nucleation on simulated secondary organic aerosol. Environ. Sci. Technol. 48 (3), 1675–1682.

Schmitt, C. G., Heymsfield, A. J., 2014. Observational quantification of the separation of simple and complex atmospheric ice particles. Geophys. Res. Lett. 41 (4), 1301–1307.

Schnaiter, M., Büttner, S., Möhler, O., Skrotzki, J., Vragel, M., Wagner, R., 2012. Influence of particle size and shape on the backscattering linear depolarisation ratio of small ice crystals - cloud chamber measurements in the context of contrail and cirrus microphysics. Atmos. Chem. Phys. 12 (21), 10465–10484.

Schnaiter, M., Gimmler, M., Llamas, I., Linke, C., Jäger, C., Mutschke, H., 2006. Strong spectral dependence of light absorption by organic carbon particles formed by propane combustion. Atmos. Chem. Phys. 6 (10), 2981–2990.

Schnaiter, M., Horvath, H., Möhler, O., Naumann, K.-H., Saathoff, H., Schöck, O., 2003. Uv-vis-nir spectral optical properties of soot and soot-containing aerosols. J. Aerosol Sci. 34 (10), 1421–1444.

Schnaiter, M., Järvinen, E., Vochezer, P., Abdelmonem, A., Wagner, R., Jourdan, O., Shcherbakov, V. N., Schmitt, C. G., Tricoli, U., Ulanowski, Z., Heymsfield, A. J., 2016. Cloud chamber experiments on the origin of ice crystal surface roughness in cirrus clouds. Atmos. Chem. Phys. 16, 5091–5110.

Schnaiter, M., Linke, C., Möhler, O., Naumann, K.-H., Saathoff, H., Wagner, R., Schurath, U., Wehner, B., 2005. Absorption amplification of black carbon internally mixed with secondary organic aerosol. J. Geophys. Res. Atmos. 110 (D19).

Schnitzhofer, R., Metzger, A., Breitenlechner, M., Jud, W., Heinritzi, M., De Menezes, L.-P., Duplissy, J., Guida, R., Haider, S., Kirkby, J., et al., 2014. Characterisation of organic contaminants in the cloud chamber at cern. Atmos. Meas. Tech. 7 (7), 2159–2168.

Schön, R., 2007. Entwicklung eines optischen messger ats zur erfassung der struktur von eispartikeln sowie der winkelaufgel osten streuung im sichtbaren spektralbereich. Ph.D. thesis, Ruprecht-Karls-Universit at Heidelberg.

Schön, R., Schnaiter, M., Ulanowski, Z., Schmitt, C., Benz, S., Möhler, O., Vogt, S., Wagner, R., Schurath, U., 2011. Particle habit imaging using incoherent light: a first step toward a novel instrument for cloud microphysics. Journal of Atmospheric and Oceanic Technology 28 (4), 493–512.

Schotland, R. M., Sassen, K., Stone, R., 1971. Observations by lidar of linear depolarization ratios for hydrometeors. J. Appl. Meteor. 10 (5), 1011–1017.

Schumann, U., 2005. Formation, properties and climatic effects of contrails. C. R. Phys. 6 (4), 549–565.

Schwarz, J., Gao, R., Fahey, D., Thomson, D., Watts, L., Wilson, J., Reeves, J., Darbeheshti, M., Baumgardner, D., Kok, G., et al., 2006. Single-particle measurements of midlatitude black carbon and light-scattering aerosols from the boundary layer to the lower stratosphere. J. Geophys. Res. Atmos. 111 (D16).

Shcherbakov, V., Gayet, J.-F., Baker, B., Lawson, P., 2006. Light scattering by single natural ice crystals. J. Atmos. Sci. 63 (5), 1513–1525.

Shimizu, A., Sugimoto, N., Matsui, I., Arao, K., Uno, I., Murayama, T., Kagawa, N., Aoki, K., Uchiyama, A., Yamazaki, A., 2004. Continuous observations of asian dust and other aerosols by polarization lidars in china and japan during ace-asia. J. Geophys. Res. Atmos. 109 (D19).

Shiraiwa, M., Ammann, M., Koop, T., Pöschl, U., 2011. Gas uptake and chemical aging of semisolid organic aerosol particles. Proc. Nat. Acad. Sci. USA 108 (27), 11003–11008.

Smith, H. R., Connolly, P. J., Webb, A. R., Baran, A. J., 2016. Exact and near backscattering measurements of the linear depolarisation ratio of various ice crystal habits generated in a laboratory cloud chamber. J. Quant. Spectrosc. Radiat. Transfer.

Sokolik, I. N., Toon, O. B., 1996. Direct radiative forcing by anthropogenic airborne mineral aerosols. Nature 381 (6584), 681–683.

Song, M., Liu, P., Hanna, S., Li, Y., Martin, S., Bertram, A., 2015. Relative humidity-dependent viscosities of isoprene-derived secondary organic material and atmospheric implications for isoprene-dominant forests. Atmos. Chem. Phys. 15 (9), 5145–5159.

Stegmann, P. G., Tropea, C., Järvinen, E., Schnaiter, M., 2016. Comparison of measured and computed phase functions of individual tropospheric ice crystals. J. Quant. Spectrosc. Radiat. Transfer in press.

Steinke, I., Möhler, O., Kiselev, A., Niemand, M., Saathoff, H., Schnaiter, M., Skrotzki, J., Hoose, C., Leisner, T., 2011. Ice nucleation properties of fine ash particles from the eyjafjallajökull eruption in april 2010. Atmos. Chem. Phys. 11 (24), 12945–12958.

Stephens, G. L., Tsay, S.-C., Stackhouse Jr, P. W., Flatau, P. J., 1990. The relevance of the microphysical and radiative properties of cirrus clouds to climate and climatic feedback. J. Atmos. Sci. 47 (14), 1742–1754.

Stith, J., Avallone, L., Bansemer, A., Basarab, B., Dorsi, S., Fuchs, B., Lawson, R., Rogers, D., Rutledge, S., Toohey, D., 2014. Ice particles in the upper anvil regions of midlatitude continental thunderstorms: the case for frozen-drop aggregates. Atmos. Chem. Phys. 14 (4), 1973–1985.

Stith, J. L., Dye, J. E., Bansemer, A., Heymsfield, A. J., Grainger, C. A., Petersen, W. A., Cifelli, R., 2002. Microphysical observations of tropical clouds. J. Appl. Meteor. 41 (2), 97–117.

Stith, J. L., Haggerty, J. A., Heymsfield, A., Grainger, C. A., 2004. Microphysical characteristics of tropical updrafts in clean conditions. J. Appl. Meteor. 43 (5), 779–794.

Sugimoto, N., Lee, C. H., 2006. Characteristics of dust aerosols inferred from lidar depolarization measurements at two wavelengths. Appl. Opt. 45 (28), 7468–7474.

Sugimoto, N., Uno, I., Nishikawa, M., Shimizu, A., Matsui, I., Dong, X., Chen, Y., Quan, H., 2003. Record heavy asian dust in beijing in 2002: Observations and model analysis of recent events. Geophys. Res. Lett. 30 (12).

Swietlicki, E., HANSSON, H.-C., Hämeri, K., Svenningsson, B., Massling, A., McFiggans, G., McMurry, P., Petäjä, T., Tunved, P., Gysel, M., et al., 2008. Hygroscopic properties of submicrometer atmospheric aerosol particles measured with h-tdma instruments in various environments—a review. Tellus B 60 (3), 432–469.

Takano, Y., Liou, K., 1995. Radiative transfer in cirrus clouds. part iii: Light scattering by irregular ice crystals. J. Atmos. Sci. 52 (7), 818–837.

Takano, Y., Liou, K.-N., 1989. Solar radiative transfer in cirrus clouds. part i: Single-scattering and optical properties of hexagonal ice crystals. J. Atmos. Sci. 46 (1), 3–19.

Tesche, M., Ansmann, A., Müller, D., Althausen, D., Engelmann, R., Freudenthaler, V., Groß, S., 2009. Vertically resolved separation of dust and smoke over cape verde using multiwavelength raman and polarization lidars during saharan mineral dust experiment 2008. J. Geophys. Res. Atmos. 114 (D13).

Tesche, M., Gross, S., Ansmann, A., Mueller, D., Althausen, D., Freudenthaler, V., Esselborn, M., 2011. Profiling of saharan dust and biomass-burning smoke with multiwavelength polarization raman lidar at cape verde. Tellus B 63 (4), 649–676.

Tunved, P., Hansson, H.-C., Kerminen, V.-M., Ström, J., Dal Maso, M., Lihavainen, H., Viisanen, Y., Aalto, P., Komppula, M., Kulmala, M., 2006. High natural aerosol loading over boreal forests. Science 312 (5771), 261–263.

Ulanowski, Z., Connolly, P., Flynn, M., Gallagher, M., Clarke, A., Hesse, E., 2004. Using ice crystal analogues to validate cloud ice parameter retrievals from the cpi ice spectrometer data. Sci. Tech. 1, 2.

Ulanowski, Z., Hesse, E., Kaye, P. H., Baran, A. J., 2006. Light scattering by complex ice-analogue crystals. J. Quant. Spectrosc. Radiat. Transfer 100.

Ulanowski, Z., Kaye, P., Hirst, E., Greenaway, R., 2010. Light scattering by ice particles in the earth's atmosphere and related laboratory measurements. In: 12th International Conference on Electromagnetic and Light Scattering. pp. 294–297.

Ulanowski, Z., Kaye, P. H., Hirst, E., Greenaway, R., Cotton, R. J., Hesse, E., Collier, C. T., 2014. Incidence of rough and irregular atmospheric ice particles from small ice detector 3 measurements. Atmos. Chem. Phys. 14 (3), 1649–1662.

Um, J., McFarquhar, G. M., 2007. Single-scattering properties of aggregates of bullet rosettes in cirrus. J. Appl. Meteor. Climatol. 46 (6), 757–775.

Um, J., McFarquhar, G. M., 2009. Single-scattering properties of aggregates of plates. QJR Meteorol. Soc 135 (639), 291–304.

Virtanen, A., Joutsensaari, J., Koop, T., Kannosto, J., Yli-Pirilä, P., Leskinen, J., Mäkelä, J. M., Holopainen, J. K., Pöschl, U., Kulmala, M., et al., 2010. An amorphous solid state of biogenic secondary organic aerosol particles. Nature 467 (7317), 824–827.

Virtanen, A., Kannosto, J., Kuuluvainen, H., Arffman, A., Joutsensaari, J., Saukko, E., Hao, L., Yli-Pirilä, P., Tiitta, P., Holopainen, J., et al., 2011. Bounce behavior of freshly nucleated biogenic secondary organic aerosol particles. Atmos. Chem. Phys 11 (16), 8759–8766.

Vochezer, P., 2015. In situ charakterisierung von kleinen wolkenpartikeln in mischphasenwolken. Ph.D. thesis, University of Heidelberg.

Vochezer, P., Järvinen, E., Wagner, R., Kupiszewski, P., Leisner, T., Schnaiter, M., 2016. In situ characterization of mixed phase clouds using the small ice detector and the particle phase discriminator. Atmos. Meas. Tech. 9 (1), 159–177.

Voigt, C., Schumann, U., Jurkat, T., Schäuble, D., Schlager, H., Petzold, A., Gayet, J.-F., Krämer, M., Schneider, J., Borrmann, S., et al., 2010. In-situ observations of young contrails–overview and selected results from the concert campaign. Atmos. Chem. Phys. 10 (18), 9039–9056.

Voigt, C., Schumann, U., Minikin, A., Abdelmonem, A., Afchine, A., Borrmann, S., Boettcher, M., Buchholz, B., Bugliaro, L., Costa, A., Curtius, J., Dollner, M.,

Dörnbrack, A., Dreiling, V., Ebert, V., Ehrlich, A., Fix, A., Forster, L., Frank, F., Fütterer, D., Giez, A., Graf, K., Grooß, J.-U., Groß, S., Heinold, B., Hüneke, T., Järvinen, E., Jurkat, T., Kaufmann, S., Kenntner, M., Klingebiel, M., Klimach, T., Kohl, R., Krämer, M., Krisna, T. C., Luebke, A., Mayer, B., Mertes, S., Molleker, S., Petzold, A., Pfeilsticker, K., Port, M., Rapp, M., Reutter, P., Rolf, C., Rose, D., Sauer, D., Schäfler, A., Schlage, R., Schnaiter, M., Schneider, J., Spelten, N., Spichtinger, P., Stock, P., Weigel, R., Weinzierl, B., Wendisch, M., Werner, F., Wernli, H., Wirth, M., Zahn, A., Ziereis, H., Zöger, M., 2016. Ml-cirrus - the airborne experiment on natural cirrus and contrail cirrus with the high-altitude long-range research aircraft halo. BAMS-D-15-00213, submitted.

Voigtländer, J., Duplissy, J., Rondo, L., Kürten, A., Stratmann, F., 2012. Numerical simulations of mixing conditions and aerosol dynamics in the cern cloud chamber. Atmos. Chem. Phys. 12 (4), 2205–2214.

Wagner, R., Ajtai, T., Kandler, K., Lieke, K., Linke, C., Müller, T., Schnaiter, M., Vragel, M., 2012. Complex refractive indices of saharan dust samples at visible and near uv wavelengths: a laboratory study. Atmos. Chem. Phys. 12 (5), 2491–2512.

Wagner, R., Linke, C., Naumann, K.-H., Schnaiter, M., Vragel, M., Gangl, M., Horvath, H., 2009. A review of optical measurements at the aerosol and cloud chamber aida. J. Quant. Spectrosc. Radiat. Transfer 110 (11), 930–949.

Wagner, R., Möhler, O., Saathoff, H., Schnaiter, M., Leisner, T., 2011. New cloud chamber experiments on the heterogeneous ice nucleation ability of oxalic acid in the immersion mode. Atmos. Chem. Phys. 11 (5), 2083–2110.

Wandinger, U., Müller, D., Böckmann, C., Althausen, D., Matthias, V., Bösenberg, J., Weiß, V., Fiebig, M., Wendisch, M., Stohl, A., et al., 2002. Optical and microphysical characterization of biomass-burning and industrial-pollution aerosols from-multiwavelength lidar and aircraft measurements. J. Geophys. Res. Atmos. 107 (D21).

Wang, B., Lambe, A. T., Massoli, P., Onasch, T. B., Davidovits, P., Worsnop, D. R., Knopf, D. A., 2012. The deposition ice nucleation and immersion freezing potential of amorphous secondary organic aerosol: Pathways for ice and mixed-phase cloud formation. J. Geophys. Res. Atmos. 117 (D16), DOI:10.1029/2012JD018063.

Wang, B., O'Brien, R. E., Kelly, S. T., Shilling, J. E., Moffet, R. C., Gilles, M. K., Laskin, A., 2014. Reactivity of liquid and semisolid secondary organic carbon with chloride and nitrate in atmospheric aerosols. J. Phys. Chem. A 119 (19), 4498–4508.

Waterman, P., 1965. Matrix formulation of electromagnetic scattering. Proc. IEEE 53 (8), 805–812.

Weinzierl, B., Petzold, A., Esselborn, M., Wirth, M., Rasp, K., Kandler, K., Schütz, L., Koepke, P., Fiebig, M., 2009. Airborne measurements of dust layer properties, particle size distribution and mixing state of saharan dust during samum 2006. Tellus B 61 (1), 96–117.

Wendisch, M., Brenguier, J.-L., 2013. Airborne measurements for environmental research: methods and instruments. John Wiley & Sons.

Wendisch, M., Müller, D., Schell, D., Heintzenberg, J., 2001. An airborne spectral albedometer with active horizontal stabilization. J. Armos. Ocean. Tech. 18 (11), 1856–1866.

Wendisch, M., Pilewskie, P., Pommier, J., Howard, S., Yang, P., Heymsfield, A. J., Schmitt, C. G., Baumgardner, D., Mayer, B., 2005. Impact of cirrus crystal shape on solar spectral irradiance: A case study for subtropical cirrus. J. Geophys. Res. Atmos. 110 (D3).

Wendisch, M., Pöschl, U., Andreae, M. O., Machado, L. A., Albrecht, R., Schlager, H., Rosenfeld, D., Martin, S. T., Abdelmonem, A., Afchine, A., et al., 2016. The acridicon-chuva campaign: Studying tropical deep convective clouds and precipitation over amazonia using the new german research aircraft halo. Bull. Amer. Meteor. Soc. (2016).

Wendisch, M., Yang, P., Pilewskie, P., 2007. Effects of ice crystal habit on thermal infrared radiative properties and forcing of cirrus. J. Geophys. Res. Atmos. 112 (D8).

Wendling, P., Wendling, R., Weickmann, H. K., 1979. Scattering of solar radiation by hexagonal ice crystals. Appl. Opt. 18 (15), 2663–2671.

Wilson, T., Murray, B., Wagner, R., Möhler, O., Saathoff, H., Schnaiter, M., Skrotzki, J., Price, H., Malkin, T., Dobbie, S., et al., 2012. Glassy aerosols with a range of compositions nucleate ice heterogeneously at cirrus temperatures. Atmos. Chem. Phys. 12 (18), 8611–8632.

Yang, H., Dobbie, S., Herbert, R., Connolly, P., Gallagher, M., Ghosh, S., Al-Jumur, S. M., Clayton, J., 2012. The effect of observed vertical structure, habits, and size distributions on the solar radiative properties and cloud evolution of cirrus clouds. Q. J. Roy. Meteor. Soc. 138 (666), 1221–1232.

Yang, P., Kattawar, G. W., Hong, G., Minnis, P., Hu, Y., 2008. Uncertainties associated with the surface texture of ice particles in satellite-based retrieval of cirrus clouds—part i: Single-scattering properties of ice crystals with surface roughness. IEEE Trans. Geosci. Remote Sens. 46 (7), 1940–1947.

Yang, P., Liou, K., 1998. Single-scattering properties of complex ice crystals in terrestrial atmosphere. Contributions to atmospheric physics 71 (2), 223–248.

Yang, P., Liou, K., 2000. Finite difference time domain method for light scattering by nonspherical and inhomogeneous particles. Light Scattering by Nonspherical Particles: Theory, Measurements, and Applications 1, 174.

Yee, K. S., et al., 1966. Numerical solution of initial boundary value problems involving maxwell's equations in isotropic media. IEEE Trans. Antennas Propag. 14 (3), 302–307.

Yi, B., Yang, P., Baum, B. A., L'Ecuyer, T., Oreopoulos, L., Mlawer, E. J., Heymsfield, A. J., Liou, K.-N., 2013. Influence of ice particle surface roughening on the global cloud radiative effect. J. Atmos. Sci. 70 (9), 2794–2807.

Yli-Juuti, T., Nieminen, T., Hirsikko, A., Aalto, P., Asmi, E., Hõrrak, U., Manninen, H., Patokoski, J., Maso, M. D., Petäjä, T., et al., 2011. Growth rates of nucleation mode particles in hyytiälä during 2003- 2009: variation with particle size, season, data analysis method and ambient conditions. Atmos. Chem. Phys. 11 (24), 12865–12886.

Zhang, Y., Sanchez, M., Douet, C., Wang, Y., Bateman, A., Gong, Z., Kuwata, M., Renbaum-Wolff, L., Sato, B., Liu, P., et al., 2015. Changing shapes and implied viscosities of suspended submicron particles. Atmos. Chem. Phys. Discuss. 15 (5), 6821–6850.

Zobrist, B., Marcolli, C., Pedernera, D. A., Koop, T., 2008. Do atmospheric aerosols form glasses? Atmos. Chem. Phys. 8 (17), 5221–5244.

Zobrist, B., Soonsin, V., Luo, B. P., Krieger, U. K., Marcolli, C., Peter, T., Koop, T., 2011. Ultra-slow water diffusion in aqueous sucrose glasses. Phys. Chem. Chem. Phys. 13 (8), 3514–3526.

Publications

Publications in peer-reviewed journals submitted in the context of my PhD work:

2016

15 **Järvinen, E.**, Schnaiter, M., Mioche, G., Jourdan, O., Shcherbakov, V. N. and et al. Near-spherical Ice in Convective Clouds. *J. Aerosol Sci.*, accepted, 2016.

14 **Järvinen, E.**, Ignatius, K., Nichman, L., Kristensen, T. B., Fuchs, C. and et al. Observation of viscosity transition in α-pinene secondary organic aerosol. *Atmos. Chem. Phys.*, 16, 4423–4438, 2016.

13 **Järvinen, E.**, Kemppinen, O., Nousiainen, T., Kociok, T., Möhler, O., Leisner, T., Schnaiter, M. Laboratory investigations of mineral dust near-backscattering depolarization ratios. *J. Quant. Spectrosc. Radiat. Transfer*, in press, 2016.

12 Stegmann, P.G., Tropea, C., **Järvinen, E.**, Schnaiter, M. Comparison of measured and computed phase functions of individual tropospheric ice crystals. *J. Quant. Spectrosc. Radiat. Transfer*, in press, 2016.

11 Schnaiter, M., **Järvinen, E.**, Vochezer, P., Abdelmonem, A., Wagner, R. and et al. Cloud chamber experiments on the origin of Ice crystal surface roughness in cirrus clouds. *Atmos. Chem. Phys.*, 16, 5091–5110, 2016.

10 Voigt, C., Schumann, U., Minikin, A. and et al. ML-CIRRUS - The airborne experiment on natural cirrus and contrail cirrus with the high-altitude long-range research aircraft HALO. *Bull. Amer. Meteor. Soc.*, BAMS-D-15-00213, submitted, 2016

9 Abdelmonem, A., **Järvinen, E.**, Duft, D., Hirst, F., Vogt, S., Leisner, T. and Schnaiter, M. PHIPS-HALO: The airborne Particle Habit Imaging and Polar Scattering probe. Part I: Design and Operation, *Atmos. Meas. Tech. Discuss.*, 2016, 1–21, 2016.

8 Wendisch, M., Pöschl, U., Andreae, M. O. and et al. The ACRIDICON-CHUVA campaign: Studying tropical deep convective clouds and precipitation over Amazonia using the new German research aircraft HALO. *Bull. Amer. Meteor. Soc.*, in press, 2016

7 Vochezer, P., **Järvinen, E.**, Wagner, R., Kupiszewski, P., Leisner, T., and Schnaiter, M. In situ characterization of mixed phase clouds using the Small Ice Detector and the Particle Phase Discriminator. *Atmos. Meas. Tech.*, 9, 159–177, 2016.

6 Ignatius, K., Kristensen, T.B., **Järvinen, E.**, Nichman, L., Fuchs, C. and et al. Heterogeneous ice nucleation of viscous secondary organic aerosol produced from ozonolysis of α-pinene, *Atmos. Chem. Phys. Discuss.*, 15, 35719–35752, 2016.

5 Hiranuma, N., Möhler, O., Kulkarni, G., Schnaiter, M., Vogt, S. and et al. Development and characterization of an ice-selecting pumped counterflow virtual impactor (IS-PCVI) to study ice crystal residuals, *Atmos. Meas. Tech. Discuss.*, 2016, 1–41, 2016.

4 Nichman, L., Fuchs, C., **Järvinen, E.**, Ignatius, K., Höppel, N. F. and et al. Phase transition observations and discrimination of small cloud particles by light polarization in expansion chamber experiments. *Atmos. Chem. Phys.*, 16, 3651–3664, 2016.

3 Hoyle, C. R., Fuchs, C., **Järvinen, E.**, Saathoff, H., Dias, A. and et al. Aqueous phase oxidation of sulphur dioxide by ozone in cloud droplets. *Atmos. Chem. Phys.*, 16, 1693–1712, 2016.

2 Kontkanen, J., **Järvinen, E.**, Manninen, H. E., Lehtipalo, K., Kangasluoma and et al. High concentrations of sub-3nm clusters and frequent new particle formation observed in the Po Valley, Italy, during the PEGASOS 2012 campaign. *Atmos. Chem. Phys.*, 16, 1919–1935, 2016.

2014

1 **Järvinen, E.**, Vochezer, P., Möhler, O. and Schnaiter, M. Laboratory study of microphysical and scattering properties of corona-producing cirrus clouds. *Appl. Opt.*, 53, 7566–7575, 2014.

Conferences, Workshops and Contributions

2016

American Meteorological Society Annual Meeting 2016, *New Orleans*
talk: *In-situ observations of ice crystal habits, their surface roughness and light scattering properties during ACRIDICON-CHUVA*
award: Student award for outstanding oral presentation

ACRIDICON-CHUVA Workshop, *Ilhabela*
talk: *In-situ observations of ice crystal habits, their crystal complexity and light scattering properties during ACRIDICON-CHUVA*

Research visit at NCAR, *Boulder*
talk: *KIT instrumentation PHIPS-HALO and SID-3 and nearly spherical ice in anvil cirrus clouds and its influence to climate*

CLOUD-TRAIN Secondment, Industry visit at Droplet Measurement Technologies, *Boulder*

2015

MPIC and IPA Kolloquium der Atmosphrenwissenschaften, *Mainz*
invited talk: *Ice crystal complexity dominates the ice cloud scattering properties*

ML-CIRRUS Workshop, *Oberpfaffenhofen*
talk: *Case studies on the ice crystal complexity during ML-CIRRUS*

CLOUD Collaboration Meeting, *CERN*
talk: *AWG2 progress report*

KIT-Repräsentantin beim Nachwuchswissenschaftler-Treffen mit dem Helmholtz-Präsidenten Prof. Wiestler, *KIT*
poster: *The nature of atmospheric ice particles: aircraft studies onboard HALO and AIDA simulations*

The 26th General Assembly of the International Union of Geodesy and Geophysics, *Prague*

talk: *Nearly spherical ice in anvil cirrus clouds and its influence to climate*

The 15th Electromagnetic and Light Scattering Conference, *Leipzig*

talk: *Application of spheroidal T-matrix model to account laboratory measurements of linear and circular depolarization ratios of ice crystals and atmospherically relevant aerosol particles*

ACRIDICON-CHUVA Data Meeting, *Mainz*

talk: *Ice particle microphysics in tropical deep convective anvils*

CLOUD Collaboration Meeting, *PSI*

talk: *AWG2 (ice microphysics) progress report*

CLOUD-TRAIN Summer School, *Cascais*

2014

American Meteorological Society 14th Conference on Cloud Physics, *Boston*

talk: *Microphysics of Frozen Droplets Formed in Deep Convective Clouds*

CLOUD Collaboration Meeting, *Manchester*

talk: *CLOUD9 campaign ice experiments*

talk: *AWG2 progress report*

talk: *CLOUD8 campaign experiment overview and preliminary results*

International Conference on Atmospheric Dust, *Castellaneta Marina*

talk: *Laboratory study of linear depolarisation ratios of different soil dust samples*

CLOUD-TRAIN Secondment, University visit at University of Manchester, *Manchester*

American Meteorological Society Annual Meeting 2014, *Atlanta*

talk: *The microphysical properties of corona-producing ice clouds observed in a cloud chamber experiment*

2013

European Aerosol Conference, *Prague*

poster: *Inter-comparison of size distribution measurements in cloud expansion studies*

INUIT Summer School, *Braunfels*

CLOUD-TRAIN Summer School, *Braunfels*

Acknowledgements

Someone wise asked me to consider all the knowledge in this world as a large balloon. Then to focus on a small section on the surface. A PhD work should be a tiny protuberance trying to push through. After my three years of being a PhD student I should enlarge the balloon of knowledge at this very specific area. I should become an expert of a specific field.

At a world, where PhD students are sometimes considered as inexpensive labour or are working as a part of large experiments without having the change to ask critical questions, I have to thank my supervisors for allowing me to conduct independent research at a field that I found motivating, to ask the critical questions, and to plan my own experiments. I am grateful for the changes to discuss on an equal footing with the establish experts of my field and I am grateful to have had the change to be visible, to represent my own research under my name. One could not hope for better starting point for their career.

A PhD thesis is a culmination of work supported by fellow students, researches, friends and family members. Here, I would like to thank all those people, who have supported me throughout my PhD project. Especially I would like to acknowledge

Prof. Dr. Thomas Leisner for giving me the possibility to work at the Institute of Meteorology and Climate Research in Karlsruhe and for supporting my work.

Prof. Dr. Manfred Wendisch for being my external co-referee.

Dr. Martin Schnaiter for supervising my work and proof-reading my thesis. Thank you for introducing me to the field of optics, for the sedulous help with the construction of the SIMONE-instrument, for giving me the change to participate to the HALO project and for trusting me to conduct independent research.

Dr. Ottmar Möhler for his supervision and support with the CERN CLOUD project.

my colleagues Dr. Paul Vochezer, Dr. Emre Toprak, Thea Schiebel, Thomas Kociok, Dr. Robert Wagner, Dr. Naruki Hiranuma and Dr. Alexei Kiselev from KIT IMK-AAF for the scientific discussions.

the AIDA technical staff, Georg Scheurig, Tomasz Chudy, Steffen Vogt and Rainer Buschbacher for their valuable help.

Cristine Nast and Katharina Weixler from KIT IMK-AAF for welcoming me, when I first arrived at the institute, and for those weekly runs.

my colleagues from the HALO project: Dr. Ralf Weigel, Willi, Dr. Stephan Mertes, Dr. Johannes Schneider, Christoph Mahnke, Christiane Schulz, Sergej Molleker, Anja Costa and Armin Afchine. Thank you for the great atmosphere during ML-CIRRUS and ACRIDICON-CHUVA campaigns.

my fellow students from the CERN CLOUD experiment for creating a warm atmosphere during the CLOUD campaigns.

Karoliina Ignatius for being my friend overseas and for the gossip sessions over a glass of red wine, which helped me to preserve my sanity after weeks in CERN.

Carl Schmitt for supporting me during my visit at NCAR and for the scientific discussions as well as great times during the RICE-campaigns.

my friends back home, Essi, Tanja, Karolina, Riikka and Ida, who support me, even if working in Germany meant that we saw us only once or twice per year.

my family for their support and especially my dad for taking care of Wilma for all these three years.

lastly, I would like to thank my *Schätzli* for his love and support.